Models of Psychological Space

John Eliot

Models of Psychological Space

Psychometric, Developmental, and Experimental Approaches

With Contributions by Heinrich Stumpf

With 15 Illustrations

Springer-Verlag
New York Berlin Heidelberg
London Paris Tokyo

John Eliot
Institute for Child Study
College of Education
The University of Maryland
College Park, Maryland 20742
USA

Cover illustration by Catherine Hewitt Eliot.

Library of Congress Cataloging-in-Publication Data
Eliot, John
Models of psychological space.
Bibliography: p.
Includes index.
1. Space perception. 2. Psychometrics.
3. Psychology, Experimental. I. Title.
BF469.E42 1987 153.7'52 87-9770

Typeset by Publishers Service, Bozeman, Montana.
Softcover reprint of the hardcover 1st edition 1987

9 8 7 6 5 4 3 2 1

ISBN-13: 978-1-4612-9156-5 e-ISBN-13: 978-1-4612-4788-3
DOI: 10.1007/978-1-4612-4788-3

Dedicated to my wife Sylvia
and to my daughters Mary and Cathy

Preface

There are several reasons to be interested in the study of psychological space. First, it is of theoretical and practical significance to understand how we perceive and represent the physical world in which we live. Second, a body of literature exists to support the contention that something like an intellectual function of spatial ability exists, distinct from general ability. Third, a growing literature provides evidence that our capacity to function spatially emerges and declines over the lifespan. Fourth, performance differences on spatial tasks have convinced some psychologists that people who perform well on such problems characteristically reason differently from those who perform poorly: they may have different general interests, or be more facile on problems requiring the imaginal transformation of figural or object arrangements. Finally, a substantial body of literature indicates that many putative spatial tests add a unique variance to the prediction of performance in many everyday activities and occupations.

My purpose in this book is threefold: to explain to family and friends what it is that I have been pursuing as a research interest for twenty years; to attempt a synthesis of theory and research in keeping with the stated purpose of the institute where I work; and to provide others with an overview of a construct that has many researchable ideas. In keeping with this last purpose, I have had in mind an audience composed of college seniors or first-year graduate students. Of course, I hope that the book will be of interest to others as well!

The book is organized so that it may be used in either of two ways. Separate chapters are devoted to each of three main approaches to the study of psychological space—psychometric, developmental, and experimental—and each chapter concludes with a brief discussion of six general issues. Thus, it is possible to consider the subject either from *within* each approach, or from *across* them, in terms of any of these general issues. Sex-related differences, for example, are treated within the context of the psychometric, the developmental, and the experimental approaches, but also as a general issue that cuts across all three.

This book is not intended as a comprehensive survey of all that is known about the construct of psychological space. Such a survey would require several volumes, and would defeat the purpose of this undertaking. My selection of models, and my description of them, furthermore, does not necessarily reflect the amount of controversy or research activity that lies behind each one. For the most part, models have been included that represent a particular position or viewpoint, and that provide illustrations of different issues. Readers are strongly encouraged to follow up interesting ideas by consulting the primary sources.

My preparation of this book has been significantly aided by the thoughtful reviews of Dr. John B. Carroll (University of North Carolina), Dr. David Lohman (University of Iowa), Dr. Dan Berch (University of Cincinnati), Dr. Nora Newcombe (Temple University), and Dr. Edward Johnson (University of North Carolina). These colleagues were very helpful in pointing out serious errors in interpretation or misstatements of fact. They are not, of course, responsible for any errors that may be found in the final version.

The chapter on philosophical viewpoints was partly written by Dr. Heinrich Stumpf of Koln, Germany. I am exceedingly grateful for his patient re-education of my understanding of different philosophical ideas. Indeed, his support and encouragement have helped to make the book possible.

Finally, I should like to thank my family for forgiving my many hours of abstract preoccupation. It will be a relief when I can put something tangible in their hands!

John Eliot

Contents

1
Characteristics of Psychological Space

An important goal of science is to describe complex phenomena in manageable and understandable terms. Scientists attempt to isolate a fragment of a whole, and then try to specify its characteristics and the factors affecting it. Yet despite a long history of philosophical and scientific theorizing about psychological space, there currently exists no widely accepted description of this phenomenon. Attempts at describing it have been like trying to define flecks of mica in beach sand: when enough particles are present their existence is evident, but it is almost impossible to describe them as an entity separate from the sand. Similarly, although fragments of the construct of psychological space are currently being researched enthusiastically, it remains a difficult task to fashion a coherent overall description of the construct from such fragments. Perhaps a useful way to begin a consideration of psychological space would be to try to specify some of its characteristics.

When people are asked, "What do you think space is?" they frequently answer with the notion of the empty vastness of outer space. Less frequently, they think of large open spaces such as oceans or deserts, and occasionally someone will define it as places where different activities occur. In most instances, people seem to think the answer requires a description, so they respond with the most general one they can think of.

When I asked my undergraduate class what space is, they found it difficult to distinguish between physical and psychological space. I suggested that the physical layout or arrangement of objects in the external world is often different from the way we perceive and think about them, and that how we respond to the shapes of figures or objects might be a way of characterizing psychological space. I then asked them for additional characteristics.

From my hastily written notes, I recorded that one student felt that psychological space refers to a position in the sequence of events. She said that it is difficult to think about space without thinking about time. Another said that it refers to our awareness of the limits and positions of our bodies as we move through different surroundings, or as we participate in physical activities like dance. Still another said that psychological space is simply our ability to reason with figures, and cited various spatial tests to "prove" the correctness of his answer. A final characterization came from a student who argued that if astronomy is the study of outer space, geography the study of space on earth, and geometry the study of the space of things, then psychological space must be the study of our *responses* to the arrangement of things, and our ability to remember where they are.

When I later asked my graduate class for characteristics, it was interesting that, after a lengthy discussion among themselves, my students decided that psychological space might best be characterized in terms of a continuum that ranged from sensory awareness to complex thought. By the end of the class, most students had aligned themselves with one position or another on this continuum.

Those at the sensory end of the continuum felt that psychological space refers to a pervasive but largely unconscious awareness of relationships among objects in the physical world. However, while some argued that this awareness is a by-product of sensory integration, others felt it is a result of cumulative internalization of order and structure from the outside world.

Specifically, those in the sensory integration subgroup argued that we obtain our knowledge of the physical world through our senses; that with increasing age comes a gradual integration of the sensory systems, and psychological space is simply a reflection of this maturing integration. Thus, the apparent inability of infants and young children to judge relationships is an index of their inadequate sensory integration. These students suggested that the visual sense might serve an "executive" function in bringing about sensory integration, and pointed out that, as a rule, adults tend to see things where they hear them.

By contrast, those in the perceptual subgroup argued that while sensory integration is important, the influence of the stimulating environment is equally so. They suggested that psychological space begins as sensory awareness but, with age and experience, becomes *perceptual* awareness as we learn to operate in terms of both the relationships in different surroundings and the relationships within particular tasks or activities.

With respect to perceptual awareness of relationships in different surroundings, the perceptual subgroup thought a hunter moving through a wilderness would be a good example. The magnitude of surrounding objects, their orientations, their distances relative to one another, and the extent to which they are mobile or stationary are all external facts that enable the hunter both to recognize familiar places and to monitor his own orientation as he moves from place to place. His survival depends upon his awareness of these facts, even if he is unconscious of this dependence.

As an example of perceptual awareness of relationships within tasks or activities, when we perceive the arrangement of words, numbers, or letter-like figures on a page, we respond to the external positional quality of this symbolic information. One student noted that when she had learned multiplication, she had memorized the arrangement of numbers in the table and, even now, "reads off" the correct answer from an imaginary table—that is, the distribution of numbers within the table contributes to the way she processes or represents information to herself. This student wondered whether others might not have difficulty with mathematical tasks because they become confused by the arrangement of information, either on a page or in their heads.

Overall, students in the two sensory-perceptual groups agreed that psychological space is fundamental to a domain of behaviors that are largely unconscious or automatic unless deliberately brought to bear in the performance of a given activity. While they differed on how to distinguish between sensory and perceptual awareness, they did agree that this awareness is common to many species, and that it involves all the senses. They wondered if it would ever be possible to measure this pervasive awareness.

A second group of students rejected the notion of psychological space as either sensory integration or perceptual responses to external stimulation, feeling it equated more with bodily awareness and the coordination of physical movement. That is, it refers to our awareness of the positions and movements of our body parts relative to other body parts; of the position, orientation, and movements of our body as a whole relative to different environments; and to the properties and movements of objects in the external world. Body movements, their sensory consequences, and external stimuli become patterned through activity because of the way the body and the physical world are constructed. They suggested that the coordination of movements might be a better index of a person's psychological space than sensory integration. Thus, the often-reported sense of disorientation among the elderly might be due, in part, to their body and movement limitations.

A third group felt that psychological space refers more to the skillful performance of spatial tasks than to sensory integration or the coordination of movements. Since it is impossible to measure our awareness of the relationships among objects in different surroundings, we should focus instead upon individual differences in the skills we bring to more focused spatial tasks. After all, differences in skill proficiency are evident in many everyday activities, as well as critical to a host of occupations.

For instance, with respect to everyday activities, individual differences are evident in such tasks as estimating how much leftover food will fit into a container, the length of string needed to tie a parcel, or the distance between two familiar locations. We differ in our skill at predicting how much space remains to the bottom of a page in a typewriter, the trajectory of a tennis ball, or how a room will look after re-arranging the furniture. We differ in how well we can judge the location of a smudge on our face when looking in a mirror, whether it is possible to squeeze through an opening in a fence, or when it is safe to merge into high-speed traffic.

With respect to occupations, there have been many studies looking at the necessity of spatial skills in a large number of artistic, technical, scientific, and mathematical occupations. A surgeon must be able to visualize the shape or displacement of hidden organs; an architect to "see" a three-dimensional structure before it is built; an electrician to trace complex wiring on schematic diagrams before undertaking repairs; a tailor to estimate the size and shape of a pattern before cutting valuable cloth; and a truck driver to visualize the dimensions of his rig before turning into a crowded intersection. In all these occupations and a great many more, successful performance depends upon the exercise of a significant number of spatial skills.

As the members of this third group conceded, it is difficult to talk about spatial skills with much precision. Whereas we can be objective when using spoken language to communicate, we are usually subjective when we respond to spatial tasks in the sense that we are not necessarily aware of spatial skills when we use them. As we mature, we tend to dismiss such skills as "obvious," because we are not conscious of our dependence upon them. Indeed, we often become conscious of our skill at estimating, predicting, or judging relationships only when we are mistaken or have somehow "misplaced" ourselves and find that we are lost.

A fourth group of students felt that psychological space is more than sensory or perceptual awareness, more than a consequence of body coordination or movement, and more than a skillful response to spatial tasks. They said that it refers to a kind of conceptual experience that transcends the perceptual moment and provides direction to our behavior. Whenever we move through different surroundings we receive information from many complex, changing, and often unpredictable sources, and we process this information with imperfect sensory modalities that operate separately or together over varying time spans. With increasing experience and memory capacity, we come to depend upon a sense of continuity or expectancy about relationships between ourselves and these environments: thus, this sense evolves gradually as a conceptual network of spatial experiences. Such a conceptual network makes it possible for us to maintain our orientation in different settings and to respond skillfully to tasks requiring transformations or displacements of materials. It thereby contributes significantly to our understanding of many different forms of complex knowledge.

With respect to such contributions, members of this group argued that, just as many of our first significant encounters with the physical world were made without benefit of language, so, too, many aspects of early learning have a distinctly spatial character. In mathematics, for example, Lesh (1976), Mitchelmore (1981), Lean and Clements (1981), and Bussman (1981) have reported evidence that our understanding of such concepts as number lines, fractions, proportionality, circular functions, etc. all depend, at least initially, upon prior spatial experiences. Similarly, Simankowski and MacKnight (1971) have argued that spatial experience underlies subsequent comprehension of such concepts as wave energy transmission, chemical bonding, cell division, x-ray diffraction patterns, contour mapping, crystal structures, and the like.

Finally, students in a group at the complex-thought end of the continuum felt that our perceptual awareness of relationships and our thinking about them are independent but related aspects of human cognition. They accepted the possibility that conceptual networks contribute to other forms of complex thought, but argued that psychological space infuses and contributes to *all* forms of thought and, indeed, exists as a form of thought in its own right. While acknowledging the difficulty of teasing apart the dimensions of most forms of thought, they claimed that psychological space is too important a constituent to be reduced to a simple category of awareness, a byproduct of movement, a skill, or a contributory experience. One student cited Spencer (1862):

> We think in relations... Now this is truly the form of all thought. Now relations are of two orders: relations of sequence and relations of co-existence; of which the first is original and the other derivative. The relation of sequence is given in every change of consciousness. The relation of co-existence... becomes distinguished only when it is found that certain relations of sequence have their terms presented in consciousness in either order with equal facility. The abstract of all sequence is Time: of all co-existence is Space. Our consciousness of space is a consciousness of co-existence (p. 164).

This student argued that, regardless of whether time or space comes first in consciousness, Spencer could be interpreted as saying that all forms of thought have a spatial dimension to some degree and, further, that this dimension may permeate our perception and representation of the world without our being conscious of it.

Another student observed that if metaphoric ability requires a capacity to recognize resemblances between two seemingly disparate forms of experience, then, as Gardner argued, metaphoric ability may be one of many kinds of thought with deep spatial roots:

> When the gifted essayist Lewis Thomas draws analogies between micro-organisms and an organized human society, depicts the sky as a membrane, or describes mankind as a heap of earth, he is capturing in words a kind of resemblance that may well have occurred to him initially in spatial form. Indeed, underlying many scientific "theories" are images of wide scope: Darwin's vision of the tree of life, Freud's notion of the unconscious as a submerged iceberg, John Dalton's view of the atom as a tiny solar system, are productive figures which give rise to, or help embody, key scientific concepts (Gardner, 1983, p. 177).

Another member of this group suggested that Boulding was saying much the same thing as Gardner when he wrote:

> It is a challenge to try and think of a metaphor that is not in some sense spatial. We think of personal relationships in such terms as near or far, obscure or clear, devious or direct, exalted or commonplace, all of which have spatial connotations (Boulding, 1965, p. 1).

Someone else pointed out that Jaynes (1977), mentioned our spatialization of our unconscious through the use of such terms as "I searched my memory," "He was lost in thought," or "It was buried deep in my memory."

Although members of this fifth group claimed that psychological space is a fundamental dimension of thought, they were frustrated by the difficulty of providing many examples. They were convinced, however, that if psychologists were to accept this position more widely, the construct of psychological space would be accorded a new status in the study of human cognition.

While I do not think that my students' responses about the characteristics of psychological space begin to encompass all possible responses, I have described them at some length because I feel they provide us with an initial description of the construct. Indeed, there seem to be at least nine characterizations that are worth serious consideration. These include psychological space as 1) a phenomenon that is common to humans as well as many other species; 2) a mental basis for a domain of behaviors that are largely automatic and unconscious at different levels of knowing; 3) a multimodal form of representation despite the domination of the visual system; 4) a form of behavior which, especially in its development, is intimately associated with our awareness of the limits and positions of our bodies as we move through different surroundings; 5) a form of symbolic processing that frequently involves imagery; 6) an expression of intelligence that contributes to our understanding of many different kinds of complex knowledge; 7) a skillful set of responses to spatial tasks, many of which are important in everyday activities and to various occupations; 8) a possible dimension in many kinds of thought; and 9) a pervasive cognitive phenomenon that is difficult to measure.

If the construct of psychological space refers to such an amalgam of physical capacities, mental processes, learned skills, forms of representation, and dimensions of thought at different levels of awareness for different tasks in changing surroundings, then how can we ever know whether a "spatial" behavior is an expression of sensory integration, movement experience, task demand, or environmental expectancy? Why is it that such an "obvious" construct is so difficult to describe?

In my opinion, there are three reasons why the clarification of this construct has been so difficult. First, we do not possess a language system that permits us to represent and communicate the complex interplay of mental and physical variables. Our present systems have made it possible to examine fragments of the construct in some detail, but not to relate them in a coherent whole. Second, our present measures of spatial behavior are useful only for the assessment of a very limited range of spatial phenomena. We need to give serious thought to the development of a new generation of instruments or procedures. And finally, we lack a construct which accounts for both our awareness of the relational distribution of things and our use of this awareness to solve problems. We need to work toward the reformulation of the construct, one that would take advantage of a new language system and would provide meaning to the results from new measures of spatial behaviors.

The creation of a language system that would enable us to represent and communicate the complex interplay of mental and physical variables is a challenge facing all aspects of cognitive study, not just that of psychological space. At the present time we have no such system, although recent work in programming computers for the parallel processing of data may be an important first step. Given the elusive nature of psychological space, we need a system that would make it possible to isolate "spatial" phenomena from other cognitive aspects, and would enable us to specify their characteristics and the factors affecting them.

Apart from computer programs, perhaps the closest analogy to the needed language system is a conductor's musical score. Most scores for large-scale pieces often contain, on the same page, the music for many different instruments. At any given moment some instruments are playing while others are silent; some are carrying the main theme while others are supporting or impeding the expression of that theme. The overall result is usually music with both a structural coherence and a sense of direction.

In the absence of an adequate language system, we are in the position of trying to concentrate enough flecks of mica from beach sand to constitute enough of an entity that can be set apart from others. An example of such an effort is Gardner's (1983) list of criteria for judging the separateness of spatial intelligence from other domains of intelligence.

According to Gardner, spatial intelligence is 1) affected by damage to specific areas of the brain; 2) evident in idiot savants, prodigies, and other exceptional persons; 3) identifiable in terms of comparatively stable (although controversial) sets of abilities or operations; 4) a domain for which a distinctive developmental history has been described; 5) a domain with an impressive psychometric literature; 6) a domain in which the interaction of task characteristics and processing solutions has been studied experimentally; and 7) a domain in which information may be encoded in one or another symbolic system for representation and communication. Spatial intelligence, for Gardner, meets all these criteria, although he admits that the evidence is not uniformly strong.

In addition to a language system, we need to give thought to the representativeness of existing measures of spatial phenomena, and to the need for a new generation of instruments that would make it possible to assess a variety of spatial behaviors in different contexts. At the present time, our existing tests measure a very limited range of behaviors.

Several years ago, I undertook to collect all the commercial, out-of-print subtests, and experimental spatial tests I could find, and by 1980 had accumulated several hundred paper-and-pencil tests from a large number of military, industrial, and academic sources. When I set out to organize this collection, I noticed that, except for differences in format, instructions, or response possibilities, all but a handful could be classified into 10 single-task and three multi-task categories. For example, the test stimuli and requirements for the geometrical construction, shape-fitting, pattern assembly, shape-completion, squares, and shape-dissection tests all appeared to be similar to paper formboard tests which require subjects to form a complete figure within a two-dimensional plane. That is, all required the mental combination of fragmented parts into a completed whole. Similarly, the form recognition test of the NFER Spatial Test 1, Thurstone's Concealed Figures Test, the hidden figure subtest of NFER's Visual Task 101, and Witkin's Embedded Figures Test all appeared to be adaptations or variations of Gottschaldt's original figure-ground perception task.

Once several hundred figural tests had been classified into task categories, I wondered whether it would be possible to arrange these categories in some meaningful way. Kelley's (1928) distinction between a factor involving "the sensing and retention of visual forms," which differed from a factor involving "the mental manipulation of shapes" provided an initial basis for the division of categories. Thus, copying tasks and figure-ground tasks, for example, were assigned to a *recognition* division, while paper-folding and surface development tasks were assigned to a *manipulation* division.

With further examination, I realized that if persons respond differently to stimulus information presented either *within* or *across* a two-dimensional plane, then it would be possible to add more categories to each division. Paper form-board tasks in which subjects form a complete figure within a two-dimensional plane, would be assigned to the recognition division, while block rotation tasks, requiring the mental manipulation of blocks across a two-dimensional plane, would be put in the manipulation division.

Moreover, I reasoned that if the ability to manipulate shapes depends upon the prior sensing and retention of those shapes, then task categories in the manipulative division would be more complex than those in the recognition division. This provided me with the basis for a rough progression of increasing task complexity. Figure 1.1 presents this logical classification in terms of the general task description for each task category, and gives a sample or two of the tasks found within each subcategory.

This classification is a logical effort at an initial taxonomy of paper-and-pencil spatial tests, not a statistical ordering of task categories. Thus, successful performance on more "complex" tasks (i.e., manipulative) does not mean that subjects necessarily will perform more successfully on "simple" tasks in the recognition division. Items from a test in the recognition division may, in fact, be more difficult to solve than some items in the manipulative division. The classification is useful, however, to the extent that it can accommodate a large number of diverse tests; to the extent that these tests can be assigned to a limited number of task categories; and, perhaps most importantly, to the extent that it underscores the comparatively limited range of existing paper-and-pencil spatial tests.

Even when one adds up the total of object or performance tests (e.g., Kohs' Blocks) and recent efforts at assessing large-scale spatial representation, the limited range of existing instruments raises serious questions about the adequacy of their representing such a pervasive and complex domain of behavior. The situation becomes more confusing when researchers administer a single test with one type of spatial item and then make broad pronouncements about spatial behavior.

In addition to a language system and more representative spatial tests, we need to work toward the generation of new models of mind, in general, and the reformulation of the construct of psychological space, in particular. Although psychology has made advances with new models of mind (Kessen, 1966; Miller & Johnson-Laird, 1976; Olson & Bialystok, 1983), much of the thinking about psychological space still centers upon Newton's ideas of an infinite, continuous, stationary, three-dimensional box or framework (Miller & Johnson-Laird, 1976). Despite recent advances in "superstring" theory, in which gravity is defined in a world expanded in nine spatial dimensions and time,

> poor dear psychology, of course, has never got far beyond the stage of medieval physics, except in statistics, where mathematicians have enabled it to spin out the correlation of trivialities into endless refinements (Broad, 1933, p. 476).

When considering the reformulation of the construct of psychological space, it is important to keep in mind that superstring theory did not spring fullblown from

Recognition Division

***Category 1:* Copying and maze tasks.** *General task description:* Subjects copy a figure superimposed upon a framework of dots or crosses upon a similar but empty framework. Maze tasks are also included in this category.

Example 1: Copying task: Instructions: Copy the figure A onto the framework provided in B.

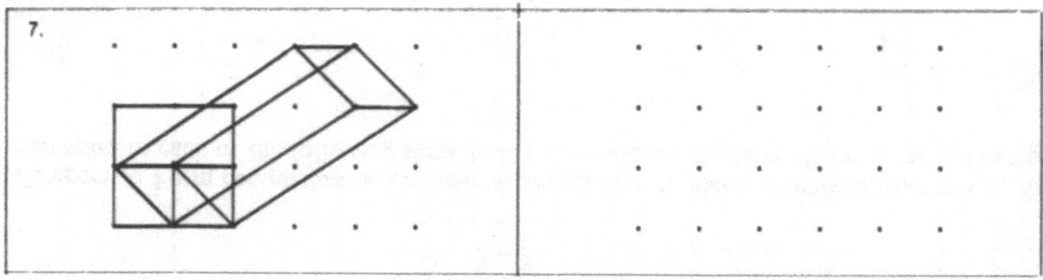

Example 2: Maze-tracing task: Instructions: Trace through maze without lifting your pencil.

***Category 2:* Embedded figure tasks.** *General task description:* Subjects identify or draw a given simple figure which is embedded, concealed, or hidden in a more complex figure.

Example 1: Hidden figures task: Instructions: How many triangles can you count in each of these figures? (From: James' Tests of Practical Ability, 1952)

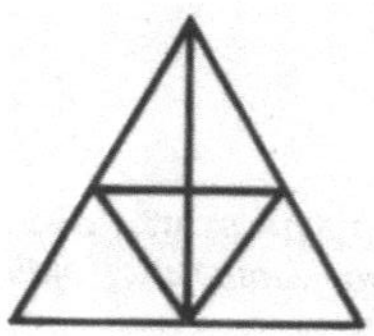

Example 2: Embedded figure task: Instructions: Which of the five figures on the right are found in the complex figure on the left? (From: SMSG: ELMA Tests, 1971)

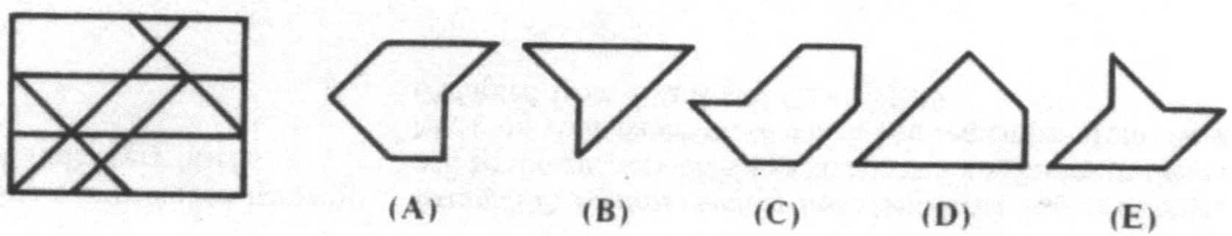

FIGURE 1.1a. Eliot's classification of figural spatial tests. (Reproduced from Eliot, J. (1980). Classification of figural spatial tests. *Perceptual and Motor Skills*, *51*, 849.)

Category 3: **Visual memory tasks:** *General description:* Subjects are shown a figure briefly and must identify or draw the figure from memory.

Example 1: Memory for designs task: Space is provided below to draw two figures which will be shown on slides very briefly.

Example 2: Figural memory task: Below is a display of objects. You will be shown, very briefly, a slide containing many of these objects. Mark on your paper those which you remember from the slide. (Adapted from Evans and Castle, 1919)

Category 4: **Form completion tasks:** *General description:* Subjects combine imaginatively the various parts to form a whole figure.
Example: In each of the following show how you would cut the large shape at the right so as to make the three smaller shapes at the left.

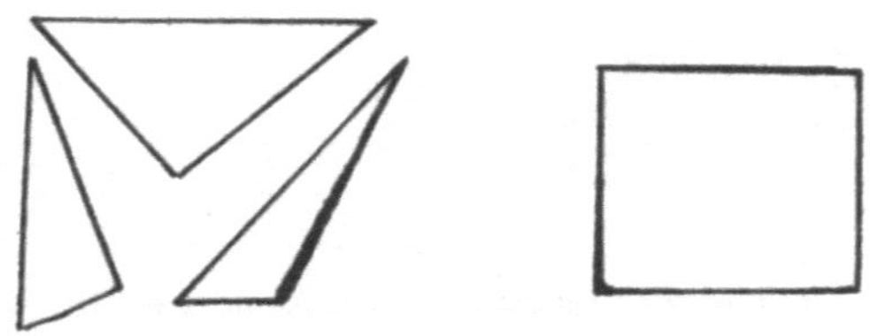

Manipulation Division

Category 5: **Form rotation task:** *General description:* Subjects indicate which of several figures, when turned or rotated imaginatively, will be the same as the given figure.

Example: Right and left hands task: Instructions: Under each item, indicate whether it is a left or right hand.

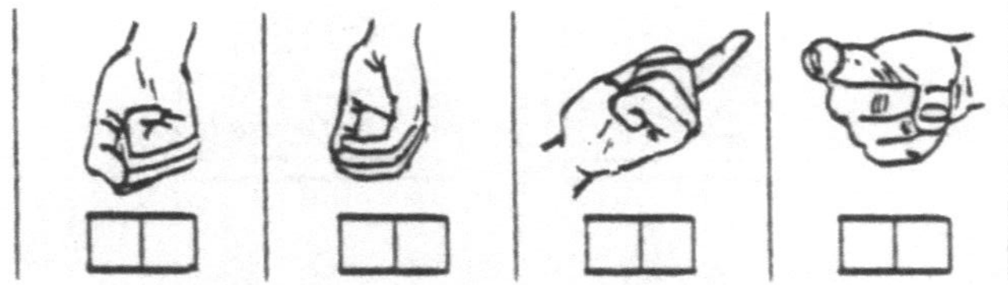

Category 6: **Blocks tasks:** *General description:* Subjects estimate number, shape, or intersection of blocks in a pile of blocks.

Example 1: Intersection task: Which of the numbered planes on the right is the shape of the intersection? (Adapted from CEEB Special Aptitude Test, 1938)

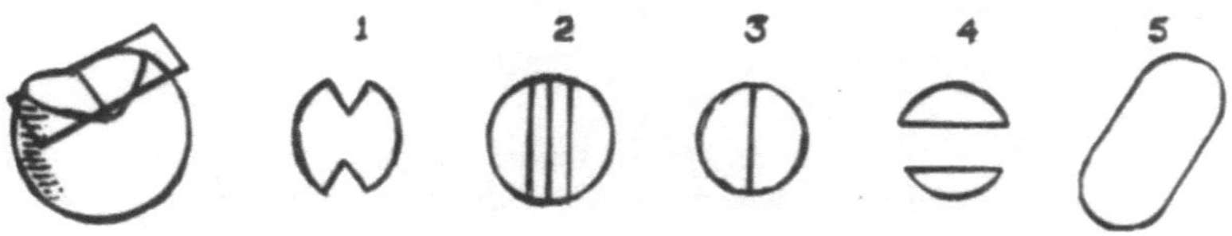

Example 2: Block counting task: Count the number of blocks in the pile below. All blocks are of the same size. (Adapted from CEEB Special Aptitude Test, 1930)

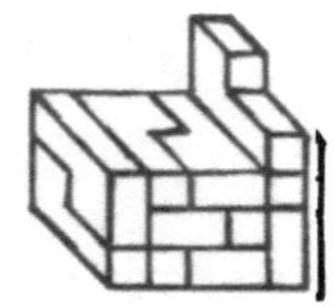

Category 7: **Block rotation tasks:** *General description:* Subjects indicate which block, when turned or rotated imaginatively, is the same as a given block or object.

Example: Identical blocks task: Which of the blocks on the right, when turned, are the same as the one on the left? (Adapted from Thurstone MATS, 1949)

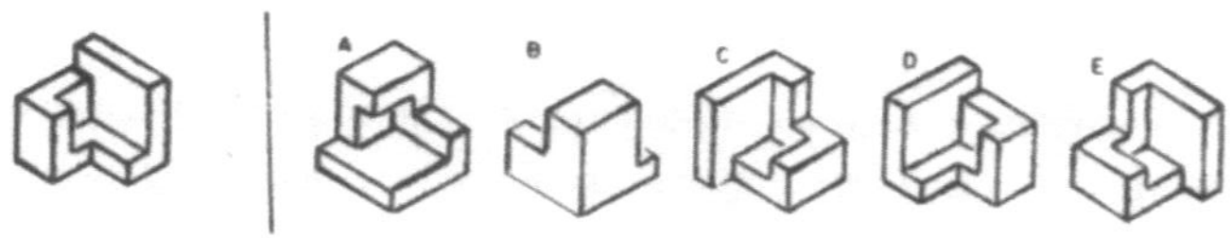

FIGURE 1.1c.

Category 8: **Paper folding tasks:** *General description:* Subjects predict mark or hole pattern of unfolded paper.

Example: Below is a drawing of two stages of a paper being folded. Indicate which of the figures on the right shows where the holes are when the paper is unfolded.

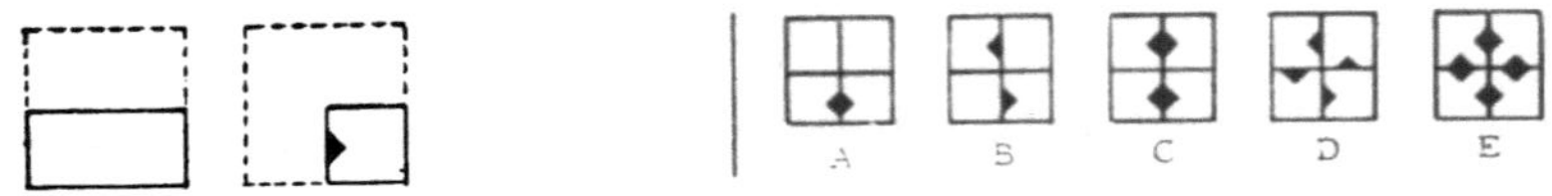

Category 9: **Surface development tasks:** *General description:* Subjects imagine how a pattern can be abstracted or formed into a given figure.

Example: Surface development or pattern comprehension task: Which of the four cubes shown could be formed by folding the pattern at the left? (Adapted from Technical Test Battery ST7, 1979)

Category 10: **Perspective tasks:** *General description:* Subjects make judgments about the arrangement of objects from different points of view.

Example: Block orientation task: The following item consists of two pictures. The left-hand picture contains twisted wire from the front of the cube. The cube in the right-hand picture is from a different perspective. Is the right-hand picture from the right, left, above, below, or behind? (Adapted from Stumpf-Fay Cube Test)

FIGURE 1.1d.

the mind of a single theorist: it required parallel developments in both physics and mathematics, and it came about through the unification of Einstein's general theory of relativity and quantum mechanics (Green, 1986). These facts are noteworthy because, similarly, it is unlikely that any reformulation of the construct of psychological space will occur in isolation.

At the present time, the literature on psychological space contains two dozen or so theoretical models and an enormous amount of empirical research. Much of this literature can be grouped and presented within either a *psychometric*, a *developmental*, or an *experimental* approach. As Pellegrino, Alderton and Shute (1984), Siegler and Richards (1985) and others have noted, these three approaches differ from one another in their assumptions about the nature of psychological space, in their preferred methods of studying spatial problems, and in the kinds of measures used to assess different aspects of spatial intelligence, spatial development, and spatial processing. As Pellegrino, Alderton and Shute have pointed out, one way to differentiate between these three approaches is to examine how each regards *individual differences* in the performance of spatial tests.

Generally speaking, the psychometricians study individual differences by administering a variety of reliable tests to large numbers of subjects under controlled testing conditions. Scores are then analyzed by factor analysis or by multidimensional scaling in order to extract the underlying factors or dimensions responsible for any patterns of correlations. Individual differences are interpreted in terms of sets of independent or partially related abilities or factors that differentiate both tests and subjects.

The developmentalists, by contrast, present specific tasks to individuals or to groups of subjects at different ages, and then examine their responses in terms of hypothesized stages of development. Individual differences are typically interpreted in terms of the achievement of cognitive structures that enable individuals to organize, represent, and transform relationships.

The experimentalists study individual differences in terms of the sensory bases of spatial perception, imagery and its relation to representation, and computational models of information processing. For the most part, differences are interpreted in terms of the means of response times or response accuracy on tasks requiring the recognition, discrimination, and generalization of stimuli under controlled conditions.

These are very general characterizations of the three major approaches to the study of psychological space, and will be elaborated upon in the following chapters. Although it should be understood that the distinctions between these approaches are not rigid boundaries, and research methods overlap (e.g., experimentalists may use instruments developed by psychometricians, etc.), they are nonetheless useful ways of organizing or concentrating a large proportion of what is known.

The concern of this book is primarily with the form or structure of psychological space: not its poetic or mythical conceptions, the values assigned to it by different cultures at different times, or the subjective, evaluative responses to surroundings as in studies of territoriality or personal space. More specifically, our purpose is to take a step toward reformulating the construct of psychological space by bringing together in one volume a sampling of theoretical models from the three above-mentioned approaches. Although we currently lack a language system for representing and communicating the complexity of this construct, as well as instruments or procedures for measuring its full range, we nonetheless possess a variety of theoretical models that have been helpful to researchers interested in different fragments of the construct. I fully anticipate that facets of these models will eventually be incorporated into a much richer construct, much as Einstein's general theory of relativity and quantum mechanics have been incorporated into superstring theory. To that end, I expect this richer construct to turn out to be much more significant than the present sum of existing models and empirical research.

To open the discussion of models from each of the three approaches, we begin with a chapter about philosophical thoughts about physical and psychological space. In the concluding chapter, I will attempt to bring together some recurring themes, issues, and questions.

As the focus of this book is on theoretical models, it is important to clarify how the term "model" is used here. For our purposes, a theoretical model is defined as a description or analogy that helps us to visualize something that cannot be observed directly (Webster's, 1963, p. 544). Reese and Overton (1970) observed that empirical theories differ from theoretical models in their comprehensiveness, and to the extent that their postulated categories and relationships provide useful, interesting, and above all testable hypotheses. Piaget, for example, is known for both his theory of intelligence and for his developmental account of spatial conceptualization. Models may thus be substructures of theories, or they may stand alone. They are useful to the extent that they draw attention to variables and set us off on new paths of research (Kosslyn, 1983, p. 114).

Both theories and models are presented as models in this book; not to denigrate theorists, but to emphasize the incompleteness of many theories (Spiker, 1966; Wohlwill, 1973). Moreover, although some early models have subsequently been outdistanced by advances in methodology and measurement, they are included here because they draw our attention to important variables, or because they have been significant in prompting or guiding research.

The descriptions of models are supplemented by references to both general and specific issues. The effect of stimulus dimensionality of test content, for example, is treated as a general issue that cuts across all three approaches. By contrast, the criterion for deciding how many factors to extract from a correlational matrix is treated as an issue specific to the psychometric approach. The general issues common to the three approaches are: 1) the emergence and decline of spatial functioning; 2) sex-related differences in spatial behavior; 3) the trainability of spatial behavior; 4) the role of imagery in spatial thinking; 5) the types of processing required by different spatial tasks; and 6) the effect of stimulus dimensionality of test content upon performance on figural, object, and large-scale spatial tasks. Each of these general issues has been or could be the subject of a separate book. My purpose in discussing them is not so much to review empirical literature as to compare the interpretations that have come out of different approaches.

Although it is tempting to begin immediately with a description of the models, such descriptions are probably more meaningful if set against a background of philosophical thought about physical and psychological space. Without going into manifold details, my intention here is to sketch in broad strokes the philosophical background from which many of the models of psychological space have emerged.

2
Some Philosophical Ideas About Space*

Although the concern of this book is with modern theoretical models of psychological space, we shall begin by tracing characterizations of space historically from philosophical through physical to psychological conceptions of space. The object of this brief review is not to compete with several existing summaries (e.g., Boring, 1942; Murphy, 1948; Jammer, 1954; Allport, 1955; Hochberg, 1962; Pastore, 1971; and Gosztonyi, 1976, among others), but to set the stage for the description of current models that in part reflect earlier thoughts. In some instances, it is very instructive to see how certain early philosophical characterizations foreshadow the different models of psychological space today.

It must be emphasized that most philosophical theories were formulated in response to critical problems of a very general nature, concerns about physical (and psychological) space often being secondary to the major arguments. One cannot describe particular theorists' philosophical concepts of space, then, without making some allusions to their major arguments.

Space as an object of philosophical inquiry can be traced to the early periods of Greek philosophy, where a dominant idea was the conception of a single unitary world (cosmos; which also means "structured order"). For most—but not all—of the early Greek philosophers, it made no sense to speak of a reality outside this cosmos. Cosmos and space were considered to be a spherical unity perceived only from the inside, from a geocentric or even homocentric point of view. The importance of this conception rested on the assumption that the whole universe was incorporated in one all-embracing or unified space (Gosztonyi, 1976, pp. 55–57). This means that, however space is described in detail, there is only one space, not several "spaces" independent or "outside" of each other. This conception of the *unity of space* persisted as a fundamental idea of philosophical theories for many centuries.

Other aspects of the classical conception of space, however, provoked controversies. For example, one important point of discussion centered on the problem of whether space was finite or infinite. Historically, the concept of infinity was

*This chapter was co-authored with Dr. Heinrich Stumpf of Köln, West Germany.

introduced by Anaximandros (610–547 B.C.) who, interestingly enough, did not apply it directly to space as such, but rather to matter. Since Anaximandros conceived of matter as having no limits, more than one world or universe could emerge out of it, with our own world or universe being only one of them. Thus, the possibility of viewing our world from the *outside* was raised. Space, still conceived as a unity, nonetheless, could contain an infinite number of worlds like our own (Gosztonyi, 1976, pp. 60–61). An important conclusion from this line of reasoning was that the borders of our world did not necessarily coincide with the borders of space.

At approximately the same time, Archytas argued even from the more traditional "inside" viewpoint that space itself must be unlimited. He asserted that place (topos) differed from matter and was independent of it. Archytas is remembered as arguing that it would be possible at the end of the world to stretch out one's hand; it thus must be possible to proceed and to stretch out one's hand again, etc. Thus, Archytas argued, one must agree that space is unlimited (Gosztonyi, 1976, pp. 63–64). A similar argument later reoccurred in various contexts as, for example, in the Atomist Lucretius, in Giordano Bruno, and in John Locke (Jammer, 1954, pp. 10–11; see below).

Another controversy arose from the fact that in the early cosmic conception, space was regarded as entirely filled with matter. The distinction between space and matter was unclear. While Anaximandros and Archytas were concerned with infinity, the Atomists concentrated on the problem of matter. On the one hand, Democritus (c. 460 B.C.) and Leucippus (c. 420 B.C.) argued that an important distinction existed between space and matter. According to them, matter consisted of atoms which, in themselves, were indivisible, unchangeable, eternal, and infinite in number. The elements differed in the form or shape that their respective atoms possess. Given the unchangeability of atoms, everything that occurred in the world was regarded as movements or displacements of atoms in some medium (i.e., space). On the other hand, for the atomists, the world was not completely filled with matter; they assumed the existence of empty gaps between atoms, and they characterized the nature of space as this gap (diakenon). Thus, from their conception of matter, they arrived at the notion of an empty space. Space, so to speak, was the nonbeing in between the being (the atoms), with the nonbeing interpreted not as something metaphysical, but real (namely space). Thus, the Atomists, despite their "materialistic" attitude, developed the concept of a being that was incorporeal or immaterial, but nevertheless real. For them, the contrast of the being and the nonbeing was located *within* our world, the nonbeing no longer viewed as something metaphysical outside our world. They also thought that space was infinite—a necessary assumption since there was an infinite number of atoms that could possibly move anywhere. Space made movement possible without influencing it. As might be expected, the Atomists' doctrine was a major revolution in the Greek philosophy of nature.

The later atomists (especially Epicurus and Lucretius) modified the older doctrine by asserting that space not only existed between the atoms, but was also the sum of the places occupied by the atoms (Jammer, 1954, pp. 9–10). Thus the qualities of the atoms made up the qualities in our world, space having no qualities itself.

Although this was a negative characterization, it led to the important concept of *homogeneity of space*. Space was viewed as having the same nature in all its parts, that is, as being *homogeneous* (Gosztonyi, 1976, pp. 69–73). This concept has remained important, if not undisputed, up to the modern time.

For the early Atomists, space was also *isotropic* in the sense that there was no preferred direction within it. The later Atomistic theorists (Epicurus and Lucretius), however, stressed the phenomenon of gravity, and insisted that the vertical direction was the natural orientation of space. Thus, for them space was still *homogeneous*, but *unisotropic*.

Controversies about the infinity of space, the problem of its relation to matter, the question of whether empty space (the void) existed, and some additional related topics not mentioned here, continued among the principal philosophical schools in ancient Greece. Depending on their respective viewpoints, some quite different arguments were put forward. For example, the Pythagoreans, asserted

> the existence of the void and declared that it enters into the heavens out of the limitless breath—regarding the heavens as breathing the very vacancy—which vacancy distinguished a natural object as constituting a kind of separation and division between things next to each other (Jammer, 1954, p. 7).

Anaxagoras (c. 488–428 B.C.), on the other hand, tried to disprove the possibility of the void by physical observations.

Although as far as we know no pre-Socratic philosopher formulated a theory of space in a systematic fashion (most of their views are, in fact, known to us only through the writings of Aristotle), many pre-Socratic philosophers contributed important ideas to our understanding of space, ideas that have had great influence over the centuries.

The two western philosophers who influenced our thinking most were Plato and Aristotle. In his dialogue *Timaeus*, Plato described space (*chora*) as something which, in its potential form, already existed before our world. For Plato, space is opposed to the real being, namely the *ideas* (or forms), and provides a place in which faint copies of the eternal ideas form our world of becoming and perishing. Plato thus distinguishes between the ideas, which are accessible to thinking (noesis), and the world of becoming, which enters our minds through sensation (aisthesis), and something third (triton genos), namely space, which, in its potential form, provides the possibility for the real being, the ideas, to come together with the nonbeing and thus form the things of our world of becoming. According to Plato, things do not emerge *out of space*, but only *in* space, as faint copies of the ideas. Once the things are formed, space, being eternal and indestructible as a potentiality, enters into its actual form—the spatiality of things. Thus, the process of becoming in our world can (in part) be characterized as a spatialization of ideas. According to Plato, there is just one idea for each species of things that we know (one idea of "horse," one of "table" etc.) and by entering into space, many different but imperfect copies of each idea arise.

With respect to our knowledge of space, Plato made an important point. Both in its potential and its actual form, space is not directly accessible to our perception. However, its properties can be inferred by "indirect conclusions" from the properties of the objects since the spatiality of objects relies on potential space. Plato argued that we can only infer space from the spatiality of the objects. He admitted that this conception might prove "an exacting demand to our willingness to believe" (Gosztonyi, 1976, pp. 77–87).

A momentous aspect of Plato's doctrine with respect to later theories is implicit in his notion of the elements, which incorporated Pythagorean as well as Atomistic thoughts. According to Plato, the atoms are composed of triangles, and the elements differ in the number of triangles building up their respective atoms. For instance, the atoms of fire have the form of pyramids (composed of four triangles); therefore, they are peaked and can permeate many bodies. The atoms of water are composed of 20 triangles and, therefore, can roll easily (see e.g., Sachs, 1917). Although this doctrine left many questions open, it was important because it applied geometric concepts to spatial phenomena (a principle also implicit in Pythagorean theory), and thus paved the way for the geometric treatment of space, although a continuous geometrization of space was to be worked out only much later (Jammer, 1954, pp. 23–24). Plato's view also implies the concept of three-dimensionality, and it is no accident that he elsewhere deplores the unsatisfactory state of the three-dimensional geometry of his time (Politeia, VII, 528).

Aristotle, who dealt at length with problems of space both in a historic and systematic way, introduced space in two different ways. In general, Aristotle did not use the term "space" (chora), but rather "place" (topos). In the few instances, however, where Aristotle employs both terms, "chora" and "topos" are used synonymously (e.g., *Physics*, Book IV, 1, 209a; 2, 209b). It thus can be said that Aristotle's inquiries concerning place were devoted to space as well. In his *Categories*, he speaks of place (or the "where") as the fifth of 10 categories. As such, place on the one hand was a basic general property of any physical object and, on the other hand, it was a general predicate that precedes any further specification of the object and was thus part of a framework in which objects are described. A specification of the place might, for example, be "in the garden."

According to Aristotle, place is a quantity. He distinguished between discrete and continuous quantities, and argued that place is continuous. In particular, he claimed that

> the parts of solid occupy a certain space, and these have a common boundary; it follows that the parts of space also, which are occupied by the parts of the solid, have the same common boundary as the parts of the solid. Thus, not only time, but space also is a continuous quantity, for its parts have a common boundary (Aristotle, *Categories*, 5a, 8–14).

This specification leads to Aristotle's second approach to the concept of space. In his *Physics* (especially Book IV), he treated space not only as a category, but also as a physical reality. Aristotle introduced his famous definition of place by likening space to a vessel. If a vessel was a portable place, he said, then place was an immovable vessel. His final definition implied that the place of an object is the "adjacent, immovable boundary of the containing body" (*Physics*, IV, 5, 212a, 20–22).

Aristotle's definition had important consequences. For a better understanding of these consequences, it is helpful to sketch some of his cosmological views first. According to Aristotle, there was no vacuum; the whole cosmos was entirely filled with bodies or particles of bodies. Aristotle held the classic view that the cosmos was structured into spheres but, in contrast to the Pythagorean doctrine, he asserted that earth was in the center of the cosmos and was immovable. According to Aristotle, there were five elements, the classical, empedocleic ones—earth, water, air, and fire—and a fifth, ether (which therefore later was called 'quinta essentia'). Among these elements, there was a certain rank order, in which ether occupied the first place, because it is the element of the outer spheres of our cosmos that comes nearest to divine being.

As the "natural place" of ether is in the spheres of the stars and planets, the elements in our sublunar world also have their "natural places" in which they tend to move unless they are impeded by force. Aristotle argued that this becomes evident by many natural phenomena, for instance by the facts that a flame is directed upwards and dirt falls downward. It is interesting that Aristotle sometimes spoke as if the tendency of the elements to move into their natural places was due to forces inherent in space itself (e.g., *Physics*, IV, 208b).

From his definition of place and his cosmological views, it follows that space—as the place of all places or the place of the whole cosmos—was the inner border of the outer sphere. Outside the cosmos, there was no space. It further follows that space was limited and that the borders of space coincided with the borders of the cosmos. The view that the universe has an outer border and a center that also is the center of the earth further implies that there is at least one preferred direction in the universe, namely the vertical.

Strictly speaking, *space* was only the spherical border of the cosmos as *place* by definition was the surrounding border of the body. Between those borders, there was, literally, no space, but only the cosmos or the body, respectively.

Since all places of bodies are contained in and thus are relative to the spherical border of the cosmos (or space), space, in a wider sense, gets an "inner" extensional structure. Aristotle, consequently, ascribed the three extensions of length, breadth, and depth not only to the bodies, but also to space. It has been repeatedly noted, however, that this wider use of the concept of space transcends Aristotle's own definition (Gosztonyi, 1976, p. 94).

In summary, Aristotle introduced space as a category. As such, it is no substance in itself, but rather is an accident of a substance, of a physical body. By definition, place is the boundary of the body. Space can be viewed as the totality of all places or as the boundary of all bodies occupying places, and thus as the boundary of our cosmos. All places are relative to that boundary. In a somewhat wider use of the concept, Aristotle described space as having three extensions and thus as being *three-dimensional*. This implies that, according to Aristotle, space is *unisotropic*. Aristotle's doctrine of the "natural places" of the elements further implies that space is also *inhomogeneous*; Aristotle went so far as to assert that the different conditions of the various parts of space exert active powers on the elements to move them into their "natural places." Movement is only a displacement of objects within space. Space itself is *immovable*.

Aristotle's theory has been treated at some length here because it was extremely important with respect to the further development of the concept of space in western philosophy. The various points raised in the preceding summary all imply important decisions taken about the problem of space or important foundations laid for the development of later conceptions concerning space.

Space conceived as a category became important in later theories that regarded space as a basic form of our knowledge and, therefore, a basic property of the objects of our knowledge (e.g., Kant). However, before this "modern" view could develop, another decision taken by Aristotle proved to be extremely influential over the centuries, especially for medieval philosophy: Regarding space as an accident implied treating space within the basic Aristotelian category of substance and accident, as either a substance or an accident. In the Middle Ages, even those philosophers who had objections about regarding space as a mere accident (for instance, William of Occam), still treated space within the framework of substance and accident, and it was a major change to regard space as something outside this dichotomy. Because of its clarity, Aristotle's definition of place exerted a great influence on philosophical theories, although the main problems of this definition led to objections by Aristotle's direct successors, Theophrastus and Straton. Two major points of criticism were that, strictly speaking, the definition restricted space to the boundary of the body, and that Aristotle insisted that the containing body has to be everywhere in contact with the contained, and thus denied the existence of the vacuum (Jammer, 1954, pp. 18–20; Gosztonyi, 1976, pp. 106–109). Although these two problems were overcome when the Aristotelian definition was finally abandoned after the Middle Ages, a third implication of his concept of space influenced later theories as well. Aristotle's definition tied space to corporeality. The difficulty of conceiving of space as independent from matter was evident in many later theories, up to the time of Descartes.

Aristotle's view that all places are relative to the immovable boundary of the cosmos is thought to foreshadow more recent conceptions, such as Newton's theory of an absolute and a relational system of space. Although these more recent conceptions differ from Aristotle's view, some authors treat Aristotle's doctrine of the "natural places" and the forces inherent in space as being fundamental to the even more modern concept of a field of force (Jammer, 1954, pp. 17–18).

Before concluding this brief discussion of Aristotle's views, another important point in his writings should be mentioned. Aristotle appears to have been the first Greek philosopher who dealt with *spatial perception* (*De anima* II, 7; 111, 1, 425). For him, spatial perception was primarily dependent on the sense of touch, although other senses were also employed in perceiving shape and size. His view was that information from different senses was integrated in a "*common sense*" (koinon aistheterion; sensus communis) – a view that anticipated some current psychological models, in which spatial perception is regarded as a result of the common functioning of several senses, especially sight, touch, and hearing.

In this brief review of theories of space in ancient Greek philosophy, we have identified and described four prominent views, each of which provided different interpretations of space and raised important questions for later research and theorizing.

The earliest view was the classical pre-Socratic philosophy of nature, which identified the borders of the cosmos and space and, by postulating the unity of the cosmos, introduced the important idea of the *unity of space.*

The second was the Atomistic theory, which clearly distinguished matter and space and introduced space as something *immaterial*, but *real*. The Atomists also implied the importance of *homogeneity of space*, and implicitly raised the problem of whether space is to be thought of as *isotropic* or *unisotropic*. Discussions and controversies among the Atomists and the other schools began the longstanding debates about the *infinity of space* and the possibility of an *empty space* or vacuum.

The third view, Platonic theory, stressed both the distinction between the *ideal* world and our (spatial) world of becoming, and the distinction between space in its *potential* and its *actual* form (the spatiality of things). It elaborated the Pythagorean thought of an analysis of matter and space by geometric concepts, and developed the notion that we cannot perceive space directly, but can only know about it by inference.

The fourth view, Aristotle's, proved to be particularly influential because it literally dominated the greater part of medieval philosophy. It implied the view of *space as a category* (which was later reinterpreted as a form of our experience), the definition of place as the adjacent boundary of the containing body, and the interpretation of space as the totality of places, immovable and limited by the outer sphere. Aristotle's theory denied the existence of a vacuum. It implied the doctrine of "natural places" and of forces inherent in space. It assumed three-dimensionality and regarded space as both unisotropic and inhomogeneous. By regarding space as an accident, it introduced the tradition of treating space within the categorical framework of substance and accident. Aristotle's interpretation of space as dependent on the physical bodies, as finite and immovable as a whole has, by some authors, been regarded as a hindrance for the development of theorizing about space up to the end of the Middle Ages and even later (Jammer, 1954, ch. 3; Gosztonyi, 1976, p. 109). Nevertheless, it contained many points of reference to more modern views. Moreover, Aristotle was the first to contribute some systematic observations about spatial perception.

The Atomistic, Platonic, and Aristotelian conceptions of space served as the prototypes of western theories throughout the Middle Ages. Early Scholasticism was oriented toward the Platonic view. Plato's doctrine of space, however, was less and less interpreted in mathematical or geometric terms. Under the influence of thinkers like Augustine (354–340), it was more and more interpreted metaphysically and theologically as speculation became a basis for deciding the traditional problems in the theory of space. Thus, for instance, Augustine argued that space could not be infinite since this would conflict with the concept of an almighty God.

In Early Arab philosophy—also dominated by theological interests and greatly influenced by Aristotle—a few thinkers favored an Atomistic point of view. Their theory of space, however, differed from the doctrine of the Greek Atomists in that the atoms were held to have no extension and were thought of as points in space. The three-dimensionality of space and of the bodies in space was thus explained by the relations of many atoms occupying different locations (Jammer, 1954, pp. 60–64; Gosztonyi, 1976, pp. 143–153).

In Europe, the period of middle and late Scholasticism was dominated by the Aristotelian view. Again, theological doctrines played a decisive role in settling debates on the core issues of the philosophy of nature. One of the most prominent Scholastics, Thomas Aquinas (1225–1274), adopted Aristotle's definition of place as the boundary of the body as well as the larger part of Aristotle's cosmology. He added theological ideas such as the notion that God is spaceless and beyond any spatiality, but that space—like the whole cosmos—is nonetheless under his power (*Summa Theologiae*, part I, quaest. 8, 52).

It is typical of mediaeval thinking about space that wherever the traditional philosophical problems were debated, the solution was not sought in theoretical considerations or experimental observations, but in theological doctrines. At times, debates were simply concluded by authoritarian verdicts, some of which in fact contradicted Aristotelian theory. In 1277, for example, the Council of the Doctors of the Sorbonne declared the Aristotelian view that there is an absolutely immovable body in space to be heretical because this view contradicted the doctrine of an almighty God. God, the council asserted, could move the whole universe (then thought to be limited) through empty space (Gosztonyi, 1976, p. 133). This decision illustrates that,

> Although men were eager to know the truth, the values of the times were such that they thought the truth would be revealed to them in accordance with Divine Will, and they looked to dogmas to guide them (Boring, 1946, p. 7).

Nevertheless, it would be a serious mistake to underestimate the importance of mediaeval thought in the development of theories of space. Consider, for example, the work of the Silesian philosopher, Witelo (1220–1277). Witelo analyzed the sense of sight and decided that it is directed only at the perception of light and color. He noted that sight on its own can only perceive things as being located in a plane, especially when objects are distant. Thus, he thought that spatial perception could not rely on sight alone, but must involve an association and combination of sensory elements (Baeumker, 1908).

During the 14th century, Aristotelian theory of space was gradually superseded by new conceptions. This was partly due to the fact that various Scholastics (most prominently Hugh of St. Victor [1096–1141] and Nicholas of Autrecourt [died after 1350]) had tried to revive the Atomistic doctrine. Other philosophers criticized Aristotle's views directly, probably the most important among them being the Jewish thinker, Hasdai Crescas (1340–1412), who after rejecting Aristotle's definition of place, developed his own conception by integrating ideas from Aristotelian and Atomistic origin. Crescas described space as an empty dimensional extension that is capable of taking up matter. According to him, space is a unity, homogeneous, immaterial, infinite, immovable and without "inner powers." By taking up matter, certain parts of empty space become the places of physical bodies and their extension (Jammer, 1954, pp. 74–80). By insisting on the existence of the vacuum, Crescas stressed the independence of space from matter; by denying that space has "inner powers," he freed the theory of space from metaphysical and theological speculations, and by defining space as a dimensional unity, he reintroduced the idea of treating space geometrically. Although his writings were long in receiving attention in European philosophy, Crescas is thought to have contributed greatly to our modern view of physical space as being free of metaphysical forces and as being independent of substance, form, matter, and physical qualities. His theory is central to the modern idea of abstracting space from physical processes, and his view of an immovable, "absolute" space reoccurs later in Newton (Gosztonyi, 1976, pp. 196–199).

Italian and German philosophers of nature, who developed their doctrines in a parallel way during the 15th and 16th centuries, were important for the theory of space because they attempted to explain natural phenomena by natural principles, rather than by metaphysical or theological doctrines. They raised highly speculative ideas of a neo-Platonic origin, and thus arrived at views that appear modern and proved to be directive for later theorizing about space.

The Italian, Bernhardino Telesio (1508–1588) described space as a container of the being. According to Telesio, whose doctrine resembled that of Crescas, space is incorporeal, immaterial, formless, free of inherent forces, and immovable in its entirety. From this assumption, it follows that it is totally homogeneous and without qualitative differences among its parts. Telesio's assertions that vacuum is required to explain movement, but that movement is caused by physical forces and not by forces inherent in space showed that he treated space as a topic of physics rather than of metaphysical cosmology. His natural-scientific doctrine is viewed as a major turning point in the theory of space, and some of his basic concepts (mass, space) were later incorporated into Newton's thought (Jammer, 1954, p. 91; Gosztonyi, 1976, pp. 205–207).

Beside the empiricism of men like Telesio, the philosophy of the Renaissance was full of speculative ideas (Windelband & Heimsoeth, 1957, pp. 306–322). The idea of the infinity of space played a major part in these, especially in the writings of the German, Nicholas of Cusa (1401–1464) and the Italians, Franciscus Patritius (1529–1597) and Giordano Bruno (1548–1600). On the one hand, Patritius, who developed a metaphysical theory of light, is important because he explicitly refused to treat space within the traditional Aristotelian scheme of substance and accident. According to him, space exists before matter, and the study of space was fundamental to all natural science (Jammer, 1954, pp. 84–85). On the other hand, Bruno appears to have been influenced by Crescas (Jammer, 1954, p. 96), but he went far beyond empirically oriented conceptions by asserting the existence of a countless number of worlds like our own in endless space. He eventually was sentenced to death for refusing to retract these ideas (Glockner, 1968, pp. 389–394).

Although based on speculation, ideas about endless space and a multitude of worlds facilitated a major revolution in thinking about cosmic space: namely, the transition from a geocentric to a heliocentric world view. It should be noted, however, that Copernicus, although he introduced that revolution with his work, '*De revolutionibus orbium coelestium*', still assumed the existence of a system of spheres (if a heliocentric one), not of single planets. Both Copernicus and Kepler (who, in turn, criticized Bruno on many points) held a homocentric view by insisting that the earth has a special status as the home of man.

Galileo finally conceived of the planets and stars as bodies moving or resting in space. Both empiricist philosophers like Telesio and speculative ones like Nicholas of Cusa and Bruno thus can be said to have paved the way for the Copernican and Galilean revolution; the empiricists by stressing the importance of experience, observation, and empirical foundation of theories of nature, and the speculative philosophers of Renaissance by widening our view of the cosmos as containing many "worlds" like ours. Galileo, through his astronomical observations, provided empirical evidence that the speculative ideas of men like Nicholas of Cusa and Bruno were not so wrong after all (Gosztonyi, 1976, pp. 222–229).

Galileo also arrived at a conclusion about space which is evident in modern thought. He noted that there is the space in our everyday experience and our scientific observation: the space of intuition. On the other hand, he also noted that the idea of an infinite space forces us to transcend intuitive space by mathematical and geometrical construction into the concept of abstract or constructive space, which is no longer fully accessible to our senses and experience.

René Descartes (1596–1650) made further important contributions to the discussion of space. In his *Meditations*, Descartes (1641), struggling against skepticism, argued that a special subjective evidence is both certain and fundamental to knowledge. Specifically, he contended that each one of us is a "thinking matter" and therefore must exist. Knowledge is only possible if it is founded on certain ideas that can be clearly and distinctively (clare et distincte) understood (and therefore must be true). Since these ideas—and spatial extension is one of them—are fundamental to our knowledge and must be involved as our knowledge develops, Descartes conceived of them as being innate ideas (ideae innatae).

Although Descartes sometimes spoke of these ideas in a nativistic psychological sense, his primary concern when using this expression was that they are immediately evident to our reason (cf. Windelband & Heimsoeth, 1957, pp. 334–337). He held, for example, that space must be involved when the objects of the outer world (which Descartes characterized as "extended matter" [*res extensa*] in opposition to the "thinking matter" [*res cogitans*]) are perceived. An important implication of his position is that analytic geometry proceeds by *constructing* its elementary objects (like the circle, triangle, plane, etc.) out of such innate ideas as spatial extension or shape, rather than by abstracting them from physical objects (Gosztonyi, 1976, pp. 237–243).

Although many of the questions addressed by Descartes had been discussed by philosophers before, his responses to these questions served to refocus western philosophical debate on certain key problems. Two of these, which became issues of longstanding controversy, will be mentioned briefly here. The first concerns the notion of innate ideas (see below), the second is implied in Descartes' dichotomy between thinking matter and extended matter. As just outlined, Descartes believed in the existence of innate ideas, ideas that are not derived from experience, but which come to mind with such certainty that they cannot be ignored or disputed. Specifically, in the famous "wax example" of his second *Meditation*, Descartes (1641) argued that our knowledge of all qualities of a piece of wax can be doubted, up to the principle that it has spatial extension. It cannot be denied, however, that the physical objects are spatially extended. Moreover, spatial extension is a principle which cannot be broken down into several specifications by our mind; any attempt to do so would be an "idle abstraction" (Descartes, 1628, Reg. XII).

Now, on one hand, Descartes held that spatial extension is a necessary, innate idea of our mind–which itself is not spatially extended, but a thinking matter. On the other hand, he also held that it is a necessary, basic predicate of the physical world. Thus, our mind, a thinking matter, is in possession of an innate idea, which in turn reflects a basic property of something completely different from our mind, the extended matter. Obviously, the concept of space plays a key role when explaining how the thinking matter can know something about the extended matter.

Descartes' response to this problem will not be discussed in detail here because this would lead us too far astray. However, it will be remembered that in Aristotle's philosophy, the concept of space had a similar double aspect. Space was introduced as a category and, yet, it was also conceived of as a property of physical objects. In Descartes' approach, the opposition of these aspects of space was sharpened. Thus, Descartes' theory may be considered as the starting point of the longstanding debate as to whether space is best considered as an idea or as something real, that is, the discussion of ideality versus reality of space.

Two lines of development in the theory of space after Descartes will be followed here. The first one dealt with space mainly as an object of physics and culminated in Newton's *Principia* (1687) and *Optics* (1704). The second discussed space as an gnoseological concept. This development began with debates on such problems as the innate ideas and as the question of reality or ideality of this space, it later became divided into separate philosophical and psychological arguments under the influence of the British empiricists and Kant.

As an example of the first line of development, Isaac Newton's (1643–1727) conception of space will be mentioned because, first, it became fundamental for classical physics and, second, because as a theory of physical space it marks the end of a tradition in natural philosophy, which is still directly relevant for psychological theories of space. The concepts of space in modern physics that replaced Newtonian theory after it had dominated the scientific world view for about 200 years are so specialized and abstract that the psychological theories dealt with in the subsequent chapters refer to them only sporadically.

Newton's conception of the physical universe united existing knowledge in mathematics, astronomy, and physics in such a way that other facts became predictable from a few simple laws. For Newton, the physical world consisted of time, space, force, and mass, with "space made up of points, time composed of instances, and both existing independently of the bodies and events which occupy them" (Jammer, 1954, p. 97). In his *Principia*, Newton distinguished between two types of space, absolute and relative:

> Absolute space, in its own nature, without relation to anything external, remains always similar and immovable. Relative space is some moveable dimension or measure of the absolute space, which our senses determine by its position to bodies (p. 6).

Absolute space is not directly accessible to our senses, but it must necessarily exist as the receptacle of the universe and the system of reference in which all places in the universe are located. However, not only absolute space—by a similar reasoning as the one mentioned in the preceding remarks on Galileo—is inaccessible to our senses:

> Since space is homogeneous and undifferentiated, its parts are imperceptible and indistinguishable to our senses, so that sensible measures have to be substituted for them. These coordinate systems, as they are called today, are Newton's relative spaces. In modern physics coordinate systems are nothing but a useful fiction. Not so for Newton (Jammer, 1954, p. 99).

Thus, similar to the respective view in Plato, Newton argued that relative space as well is accessible to us only via the relations of the objects in it.

An example of a relative space is the space occupied by a lake on the surface of the earth. The place of the lake is (relatively) fixed with respect to the environment, but it is moving as the earth moves in absolute space. Thus, with respect to its relative space, the lake is at rest, but with respect to absolute space, it is moving.

By contrast, absolute space is conceived as immovable and homogeneous. It was conceived as a container or framework which, like the Cheshire Cat's grin, was something that remained after the contents had been removed. Newton assumed, further, that the center of absolute space was also the center of gravitation which, according to his view, must be located near the center of the sun.

With Newtonian theory, we leave the discussion of space as an object of physics and cosmology. The reader interested in the impact of Newton's theory on physics and on our modern world view as well as the more recent revisions of the concept of space in physics is referred to Jammer (1954) and Gosztonyi (1976).

With respect to the second line of development, the works of the British empiricists gave rise to a further clarification of the concept of space. As will be remembered, Descartes had introduced the notion of innate ideas to explain how our mind can come to know something about the outer world, which is totally different from it. Descartes introduced the concept of innate ideas as an instrument in philosophical discussion designed to describe the a-priori element in our knowledge; taken literally, however, this concept is also subject to a psychological interpretation, and this is just how it was taken up by the British empiricists (especially Locke [1632–1704] and Hume [1711–1776]).

The empiricist's basic attitude toward innate ideas can be summarized as follows. If there were innate ideas, then everyone would agree on them. Now, with respect to any idea which has been claimed to be innate, there are examples of people who do not know it or do not agree on it (see especially Locke, 1690, I, II). Therefore, "it is evident that there are no such impressions" (Locke, 1690, I, II, 5). Specifically, Locke likened our mind to a "white paper void of all characters" into which all ideas enter "from experience; in that all our knowledge is founded, and from that it ultimately derives itself" (Locke, 1690, II, I, 2). He went on to argue that ideas arise either from sensation or from reflection, the latter being his term for the mind's knowledge of its own operations. The generation of complex ideas from simple ones is one of the operations of the mind which reflection reveals.

The idea of space, according to Locke (1690, II, IV), is a simple one arrived at by sight and touch. Space differs from the body in that it is *without solidity* (solidity, not—as in Descartes—extension, for Locke is the primary quality of the body), *without resistance* against moving bodies, *continuous*, and *immovable* (Locke, 1690, II, XIII, 11–14). Strictly speaking, spatial perception is not directed at space itself, but at the spatiality of the bodies and at spatial relations. Because when imagining spatial relations, our mind can proceed indefinitely, space must be infinite (Locke, 1690, II, I, 24), a reasoning also implicit in Archytas, Lucretius, and Bruno, among others (Gosztonyi, 1976, p. 305).

Berkeley (1685–1753) elaborated on Locke's theory regarding the interaction of sight and touch in spatial perception, and contributed a number of important observations for a comprehensive theory of spatial perception. According to Berkeley, the impression of spatiality, provided by the sense of touch and spatial vision, is only possible through the association of the impressions of vision with impressions of touch. Berkeley's theory emphasizes that the interaction of those two senses proceeds via association of impressions, not via logical inference (although he speaks of "acts of judgment" [Berkeley, 1709, p. 11]). From our experience, we learn to associate certain optic impressions with haptic or kinaesthetic ones, and thus arrive at spatial perception. Once we have achieved the perception of spatiality of an object, the haptic or kinaesthetic impressions are no longer required, and it is sufficient that they can be recalled from our memory.

When discussing spatial vision, Berkeley (1709, p. 16, 27) also alluded to sensations of tension when we adjust the muscles of our eyes to view distant objects.

Although he does not elaborate on this matter further, he appears to have been the first to mention both the phenomenon of convergence in binocular vision and the phenomenon of accommodation (Gosztonyi, 1976, p. 313).

In later writings, Berkeley (1710) modified his original position by denying the existence of absolute physical space and by asserting that nothing exists outside our own thoughts. By so doing, he reduced Descartes' problem of how thinking matter comes to know about extended matter to the contention that extended matter exists only in the perception and the imaginations of the mind. Thus, for the outer world to exist means only that it is perceived (*esse est percipi*). The psychological treatment of traditional gnoseological problems by Locke and the further development of his theories by Berkeley thus provided an empirical foundation to several of our modern psychological theories of spatial perception.

The British empiricists claimed their theory of knowledge, later modified and formulated in an even more radical way by Hume (1739; 1748), to be valid both as a philosophical and a psychological theory. Locke's criticism of innate ideas, however, did not prevent the concept from being employed in an equivocal way, both philosophically, designating the a-priori element in our knowledge, and in a psychological sense, describing innate components in our cognitive development. As previously noted, the latter interpretation of the concept was secondary for philosophers like Descartes, but a clear distinction between the philosophical and psychological interpretations of the concept was still to be worked out.

Although the German mathematician, Lambert (1728–1777) insisted that the problem of a-priori concepts was a logical and not a psychological one (cf. Windelband Heimsoeth, 1957, p. 396), and although Leibniz (1646–1716) generally wrote of "virtually innate ideas" meaning that they were implied in the functions of mind but were not necessarily conscious in every case, the writings of Hume widened significantly the cleavage between empiricists and nativist philosophers with respect to the question of how we obtain our knowledge of the world.

Toward the end of the 18th century, Kant (1724–1804) introduced a clear distinction between the philosophical and the psychological interpretations of the doctrine of innate ideas. Moreover, he offered a philosophical solution to the debates about innate ideas as well as to the problem of the reality or ideality of space (and time). Kant began his *Critique of Pure Reason* with an empiristist statement: "There is no doubt whatsoever that all our knowledge begins with experience" (Kant, 1787, p. 1). However, he went on to add that "nevertheless, it does not necessarily all spring from experience." He then argued that we learn about the foundations of our knowledge through a long process involving experience, and that although experience has *temporal priority* in our knowledge, the bases of our knowledge have *logical a-priority* over experience, since without certain prerequisites of knowledge, our experience would be impossible. Kant's distinction between (temporal) priority and logical (or as Kant says in this context, "transcendental") a-priority can be illustrated by a mathematical proof, where the conclusion, although being logically a-posteriori, may be known to the students in advance of some premise, although this is logically a-priori. It is extremely important to note—and often ignored in the literature—that in making these distinctions, Kant is arguing philosophically about the conditions of knowledge, not psychologically. Indeed, Kant is *not* concerned with either innate elements in children's perception of their environment or with problems of how children might learn about our a-priori concepts.

For Kant, the fundamental conditions of knowledge could be grouped into two classes—the two forms of intuition (*Anschauungsformen*), time and space, and the 12 categories or pure concepts of understanding (*reine Verstandesbegriffe*), such as substance, causality, and necessity. All of our knowledge is founded on both intuition and concept. Intuition may be *pure* (consisting of nothing but its form; in this sense the forms of intuition can be also called "formal intuitions" [Kant, 1787, p. 160]) or *empirical* (containing also a sensation such as the impression of blue). When we apply categories to *pure* intuitions, a-priori knowledge arises. For example, geometric propositions arise when categories are applied to space. Similarly, arithmetic propositions arise when concepts are applied to time. However, when categories are applied to empirical intuitions, everyday or scientific experience or empirical knowledge is obtained. Whereas for Kant, time is the "form of the inner sense," space is the "form of the outer sense." Thus, space is fundamental not only for geometry, but also for our whole experience of the outer world.

To explain how our experience of the outer world arises, Kant conceived of an unordered manifold of intuitions. In applying the categories to the manifold of our intuitions, various elements of that manifold are grouped together and referred to an underlying unity. For example, when seeing a boat on a river, the observer refers a series of impressions to the boat, others to the water, still others to the opposite river bank etc. When recognizing the boat, the observer groups his intuitions of the boat together and refers them to an underlying unity (or substance). Although the impressions involved and the concepts used (e.g., the concept of a boat as a special kind of substance) are those of the observer, the experience derived is about something outside the observer. Thus, a fundamental characteristic of our experience, which proceeds as series of judgments, is that we "imagine the objects as outside ourselves and all in space. In space, their shape, magnitude, and relations to each other are determined or determinable" (Kant, 1787, p. 37).

Within his theory, Kant offers solutions to the traditional philosophical problems of space as just described. For example, his solution to the problem of whether space is ideal or real is contained in his principle that "the conditions of the possibility of experience are as well conditions of the possibility of the objects of experience" (Kant, 1787, p. 197). Space as well as time and the categories, thus, has both "transcendental ideality" and "empirical reality," namely, it is a form of our knowledge, but it is also empirically real as a part of the structure of the world we perceive.

The problem whether space is finite or infinite is addressed in Kant's discussion of the restriction of our knowledge to the realm of what is a possible object of experience. According to Kant, we only can have knowledge within the limits of possible experience. Anything beyond the scope of our forms of knowledge, such as a God outside space and time, can only be an object of relevation or belief, not of knowledge. The decision as to whether the world is finite or infinite also exceeds the limits of our possible experience, and thus Kant (1787, pp. 545–551) argues that this question must be left unanswered. The only thing that can be said about these problems within a Kantian framework is that the world is of indefinite magnitude (Wagner, 1967, pp. 114–117).

With Locke, Berkeley, and Hume (on the one hand) introducing the psychological point of view, and Kant (on the other hand) delineating the philosophical point of view against it, we have reached the point historically where psychological theories of space separate from the more general philosophical traditions and become important fields of research in their own right. Despite this development, psychological theories of space have constantly returned to conceptions of space developed within the philosophical tradition. Therefore, even for those who do not share Kant's philosophical views, some conclusions from his theory should be kept in mind. The concept of innate ideas grew out of a philosophical discussion, and was then interpreted psychologically. It was in this latter meaning that writers like Locke tried to refute it. The philosophical thought behind that concept—the notion of a-priorism—has been made quite clear by Kant, but his discussions were in general restricted to philosophical problems. His doctrine has been called a "nonempirical theory of the empirical" (Prauss, 1974); he did not mean, for instance, to give explanations of how children learn about space. Kant was an a-priorist, not a nativist. What psychological view is taken with respect to nativism or empiricism is largely indifferent to transcendental philosophy. Although the Kantian doctrine has repeatedly (and successfully) been used to derive hypotheses for psychological research (see e.g., Bussmann, 1981), it was not designed as an empirical model, and results of psychological research should not be interpreted as proving or disproving it (an error that is made all too often in psychological writings).

As far as hypotheses about psychological space are concerned, other ways of thinking can be drawn on in a fruitful way as well. A consequence of the Newtonian conception, for instance, was pointed out by Gibson (1950) as follows:

> Perceived space seemed to divide up naturally into certain geometric categories. First there was extensity in two-dimensions: the bare characteristic of space as being spread out. This corresponded to the plane of the vertical and horizontal axes in geometry. Then there was the aspect of locations in two dimensions, or the localization of points in a visual field. This corresponded to the x and y coordinates of geometry. Next there was the aspect of shape or form in the visual field. This corresponded to the abstract forms of Greek geometry. Finally, there was the aspect of depth or distance, the third dimension of space, and this corresponded to the third dimension of geometry. Extensity, shape, and distance: these were the primary constituents of space (Gibson, 1950, p. 15).

The problem of how we apprehend the Newtonian universe, in other words, became essentially the problem of how we apprehend geometric space. If empty

Euclidean space was filled with objects or points whose position could be determined by the Cartesian coordinates, then the question that confronted philosophers of the 19th century was whether every point in the visual field was intrinsically connected at birth, or whether points in the visual field were differentiated through experience and then associated with an appropriate retinal point in the eye. According to Allport (1955):

Both sides agreed that "mind" had innate capacities that transcended the senses. . .but though psychologists of both sides agreed upon "spatiality" as a capacity of mind and hence were all, to that extent, nativists, they disagreed as to how spatial significance of particular receptor patterns came about.

The nativist doctrine was that local signs were intrinsically spatial in character. The points of sensory stimulation were "directly labeled" in phenomenological experience as to their loci. Hering, for example, whose work appeared from 1861 to 1864, held that signs were attached to the visual experiences arising from each retinal point. In fact he held that there were three such signs for each point, one for each of the three dimensions. To account for spatial perception it was not necessary to bring in eye movement because the spatial aspect was immediately given.

Lotze, Helmholtz, and Wundt, all empiricists, held a different view. Lotz held that the signs which were connected with sensations, or were properties of them, were not originally endowed with a spatial character but had to acquire it. The process of acquisition involved muscular sensations and the associative combination or fusion of sensory components. Every tactile and retinal point had its local label. The aspect of sensory experience which carried the spatial quality was, for Lotze, the attribute of intensity; specifically, it was the intensity of muscular sensation, an intensity pattern for the spot touched. . .With the continuous movement of the receptor in relation to the surface, the successive local signs are experienced in proper sequence (pp. 87–88).

Of the two positions, empiricism appeared the stronger at the beginning of the present century for two reasons. First, no anatomical structures were found that were sensitive to depth, forms, distance, or objects. Second, just as the empiricists had difficulty in explaining the relationship between sensation and perception with respect to our understanding of illusions, so nativists had difficulty explaining size invariance or the fact that the distance of an object can vary without change in its perceived size (O'Keefe & Nadal, 1978, p. 33).

In 1890, William James commented upon the long history of interest in spatial perception, and suggested that it could be summarized in terms of three main theories. The first, advanced by Helmholtz and Wundt, was that spatial attributes reside in objects, not sensations, and are learned through inferences. We use clues revealed through various sensory modalities to form inferences about the nature of objects and object relationships. The second theory, favored by James, was that the senses respond directly to spatial stimuli, and that the spatial senses for vision, audition, and movement operate as distinctly differentiated modalities. We use the various sensory excitations as cues to bring a given perception into the stage of consciousness. Our senses cooperate quite successfully in giving us knowledge of the spatial characteristics of objects, such as their location, size, shape, stability, and mobility. The third theory, not favored by James, was that space is an *innate* form of all perception, neither learned nor inferred from experience.

As Hochberg (1962) describes it, the next rebuttal from the nativists came, not as an attempt to show that anatomical structures existed that were sensitive to spatial information, but rather the reverse:

> The radically new approach was Gestalt theory, which proposed that 1) the ability to perceive spatial characteristics was inborn, and that 2) spatial experiences are indeed as immediate as any of the so-called sensations, but that 3) fixed relationships between stimulus and response do not exist, and are not necessary for either (1) or (2) to be true. Gestaltists conceded that individual receptor organs could not be found for the experience of distance, but, for that matter, they could not be found for, say, red, either, or for any other experience (p. 294).

For the three men most closely associated with the Gestalt movement in the 1930s and 1940s, Wertheimer, Köhler, and Koffka, all visual perception was three-dimensional because, as Köhler (1947) insisted,

> experienced order in space is always structurally identical with a functional order in the distribution of underlying brain processes (p. 39).

The visual field was thought to correspond with or be isomorphic to an underlying excitatory brain field in its relations of order, although not necessarily identical in form (Boring, 1950, p. 615). The Gestalt concept of an underlying brain organization for all perception was an interesting psychological analogue to Kant's earlier notion of an a-priori framework imposed by intuition.

As Gibson (1950) points out, Wertheimer, Koffka, and Köhler were not as explicit about the perception of space as they were about form. For the Gestalt theorists, the perception of form was organized according to recurring tendencies or laws of segmentation, relational grouping, and simplicity: underlying brain processes were said to seek out "good form" and to impose organization and meaning upon it. By emphasizing organizational properties, Gestaltists indicated one way in which perception was more than the sum of our sensations.

Although Gestalt ideas were dismissed for a number of reasons, they nonetheless were partly responsible for the gradual shift of interest in behavior qua behavior in the 1930s and 1940s to a growing interest in mental processes beginning in the late 1950s. It is interesting to note, moreover, that as recently as 1978, O'Keefe and Nadal advanced the notion that brain organization might be an a-priori basis for perceiving spatial information. Specifically, they argued that the hippocampus provides us with an a-priori Euclidean framework for our perception of the external world. This is a closer analogue to the Kantian doctrine of a-priori forms of intuition than those of the Gestalt viewpoint. Furthermore it is supported by a body of neurophysiological evidence.

In this brief review, we began with reference to the early Greek conception of a single unitary cosmos, touched upon the controversies about the infinity of space and the relation between space and matter, described Plato's and Aristotle's different conceptions, mentioned Descartes' distinction between thinking matter and extended matter, referred to the distinction between absolute and relative space in Newton's theory, dealt briefly with the empiricism of Locke, Berkeley, and Hume, set out some of the a-priori contentions of Kant, and concluded with William James's summary and a mention of Gestalt and more recent neurophysiological ideas about psychological space.

Such a review is useful to the extent that it points to some issues that continue to affect how we think about psychological space. One such issue, of course, is the centuries-old controversy over how we obtain our knowledge of the physical world. Despite the fact that some scholars claim that the nativist-empiricist debate is meaningless today, Bower and Hilgard (1983) think differently:

> One of the most engaging issues within the theory of knowledge is the question of how concepts and knowledge arise, and what is the relation between experience and the organization of the mind. Two opposing positions on this matter are empiricism and rationalism (nativism). These have been constant combattants within the intellectual arena for centuries, and strong forms of them are still recognizable today in "scientific" psychology (p. 2).

Other issues concern whether space is to be conceived as an all-embracing unity or as a multiplicity of spaces. A third issue concerns the nature of the relationship between our knowledge of the physical world and our knowledge of psychological space. Within these latter two issues a number of other points can be related.

With respect to the early Greeks' concept of a unified space, even today we tend to think of space as all-encompassing, especially when we refer to psychological space as a domain of spatial abilities and processes that differs from other cognitive abilities. That is, we tend to treat all psychological space as if it were a convenient unitary trait, and to assume that a common "spatial" processing is involved whether we are responding to problems involving two-dimensional figures, three-dimensional objects, or large-scale environments.

The conception of a unified space took on new life when Newton advanced the idea of absolute space as being a container or framework of sorts that was independent of content. Despite Einstein's later contention that a timeless or immaterial absolute space does not exist, we continue to hold to this framework conception and to locate things in terms of three-dimensional Cartesian coordinates. We have even extended this notion of an absolute framework by arguing that each of us lives in a world of concentric layers of activities – a nested framework of relationships that extends from one's room to the home, neighborhood, city, region, and nation (Buttimer, 1976, p. 284). Moreover, many of us find convincing Piaget's notion that

> What a child eventually needs to establish. . .and does not at first possess, is a picture of space as some kind of container made up of a network of sites or subspaces, which moves from site to site, now occupying a given site, now leaving it unoccupied or empty (Flavell, 1970, p. 1016).

Newton, Kant, and Piaget notwithstanding, psychological space may entail more than just the study of locational behaviors within different frames of reference, and more than study of the relationships between figures or objects in different arrangements. Again, we are hindered in our thinking about alternatives to absolute or relative spaces by the lack of a language system that would enable us to talk about the multiplicity of spaces without endangering the identity of the construct.

With respect to the relationship between physical and psychological space, recall that whereas Descartes distinguished between extended and thinking matter, Berkeley denied the existence of physical space by insisting that nothing exists outside the realm of thoughts. Newton, like Plato, argued that physical space existed, but that it could only be experienced indirectly through our inferences about the relations between objects. According to Kant, space is directly amenable to our knowledge as a formal intuition. This range of different view clearly shows up the controversy about the nature of the relationship between physical and psychological space.

Although some philosophers continue to deny the existence of physical space, the fact remains that many concepts about it have influenced our thinking about psychological space, and a number of psychologists have contended that our understanding of the latter is obtained from our interactions with physical space. Among these, O'Keefe and Nadal (1978) insist that there are important differences between physical and psychological space. They point out that the concern of the surveyor or physicist is with a material world separate from the existence of human minds, and their measures of this physical world typically employ the use of rulers or meters to establish the distance, direction, or movement of objects. By contrast, the concern of psychologists and physiologists is with space attributable to mind, either as an intrinsic aspect of, or a highly probable result of, its normal operations. Their measures of psychological space typically include tasks that require individuals to make estimations, predictions, or judgments about the relationships among objects. Psychological space itself may take many forms:

> Included are concepts which the mind constructs on the basis of reflections on experience, abstractions from sensations, organizing principles which impose unified perceptions upon otherwise diverse sensory inputs, organized sensory arrays which derive their structure from the nature of peripheral receptors, or a set of sensations transduced by a specialized (spatial) sense organ (p. 7).

As shall become evident, the issue of the relationship between physical and psychological space will surface repeatedly in the different interpretations of space.

The nature of the relationship between physical and psychological space is closely related to continuing questions about how we come to know and use systems for measuring physical space. Generally speaking, the question is how thinking matter abstracts, constructs, or incorporates a metric awareness of extended matter.

Although much of our thinking about psychological space continues to reflect Newton's and Kant's conception of a three-dimensional Euclidean space, with recent developments in mathematics and physics, we need to question whether non-Euclidean geometries or mathematical spaces of more than three dimensions might not contribute toward a richer understanding of physical space in general, and psychological space in particular.

To summarize, this chapter has provided a brief historical review of philosophical thought about physical and psychological space, and a discussion of issues that continue to affect how we characterize or think about psychological space. Given the fragmented nature of current thinking and research about psychological space, it is important to remember that many of the research questions asked today were also asked by philosophers many years ago. In this sense, this chapter may serve as a useful background to our subsequent discussion about different theoretical models of psychological space.

In subsequent chapters, theoretical models from each of the three main approaches to the study of psychological space are described. For each approach, a summary of characteristics is given, a sampling of models are described in a historical context, and several general and specific issues are discussed. We will begin with a discussion of the psychometric approach.

3
The Psychometric Approach

The psychological study of human intelligence is primarily the concern of *psychometricians*, or differential psychologists, who assume that intelligence exists as a set of quantifiable dimensions along which people can be measured. Typically, the psychometrician studies individual differences by administering test batteries that have demonstrated reliability to large numbers of subjects under controlled conditions. Scores from these tests are then analyzed by factor analysis or multidimensional scaling to extract underlying factors or dimensions responsible for patterns of correlations. Individual differences are usually interpreted in terms of factors that differentiate both the tests and the subjects. The models advanced to account for intelligence are expressed as factor structures. There are numerous expressions of these, ranging from those that emphasize a single general ability factor (G), to those that emphasize a large number of specific abilities.

Psychological space is only one of several domains of abilities within factor structures. Psychometricians interested in this domain usually ask questions about whether "space" can be quantified, whether it exists as a unitary trait or as several unrelated abilities, whether it is associated with particular patterns of abilities, and whether useful predictions can be made from performance on spatial tests. They usually assume that in the assessment procedure subjects' performances are representative of their best efforts, the examiners' administration of tests is consistent across situations, the battery includes tests that sample different abilities adequately, and that the subjects tested are similar to those used for the standardization of the tests. Psychometricians identify tests as "interesting" if they discriminate between groups, if the items correlate with items from established or reference tests, and if they demonstrate predictive validity.

The purpose of this chapter is to describe different models and issues from the psychometric approach. To this end, various models will be described within the broader context of the history of mental testing. Although this survey will parallel the history of mental measurement, the emphasis will not be upon the history of factor analysis or multidimensional scaling per se. Thus, although some technical terminology is required for the description of particular models, the focus is on the *models* and not on the detailed definition of such terms as factor structure, oblique rotation, mapping sentence, etc. It is assumed that the interested reader will find definitions for such terms in the many books on measurement that are currently available.

Within the history of mental measurement, psychometric efforts to identify, measure, and interpret a spatial factor can be organized in terms of three phases of activity. In the first phase (1904–1938), researchers sought evidence for a spatial factor over and above a general factor of intelligence. In the second phase (1938–1961), they attempted to ascertain the extent to which spatial factors differ from one another. And in the third phase (1961–1986), there have been attempts both to designate the status of spatial abilities within different factor structures, and to examine sources of variance that affect performance on spatial tests. It must be emphasized that this classification of activity into three phases is merely intended to define general patterns of activity, and distinctions and overlaps are therefore probable.

Before we begin the description of psychometric models, we need to be reminded about the various social, intellectual, and educational events that resulted in the testing movement in the United States and Europe. As Carroll (1982) and others have noted, mental testing began at a time when the Industrial Revolution had taken a firm hold, when there was a marked increase in population (particularly in the United States through immigration), when large-scale programs were developed to educate the masses, and when the theory of evolution had been transformed into a movement known as Social Darwinism. Darwin's (1859) emphasis on individual variation, its possible genetic basis, and the fact that variation influenced survivability all served to focus attention on differences in human adaptability. The combination of these events prompted interest in ways of selecting individuals for education, industrial training, or vocational placement.

Before Darwin published his theory of evolution, Quetelet (1842) had contributed to the history of mental testing by being one of the first scholars to apply the concept of normal probability to studies of human variation. Although the concept had been conceived earlier by other mathematicians, it was Quetelet who demonstrated that with large unselected human populations, scores of a variety of tasks tended to group together at the center of a normal curve, with fewer and fewer scores found at the extremes.

Galton (1869) was impressed by Quetelet's work, and carried it further by pointing out that people appear to differ as a result of combinations of small effects which, when brought together, give rise to a normal curve. He noted that a person's score could be defined in terms of its deviation from the central tendency of such a distribution. In his *Hereditary Genius*, Galton attempted to delineate the characteristics of persons of genius, and suggested that patterns of intelligence tend to run in families. In addition to his interest in inherited individual differences, he contributed to our knowledge about measuring these differences by developing the logic of correlation as a measure of the relationship between two variables. It was also Galton who introduced such notions as the statistical median and the standard score, as well as important techniques for establishing the reliability and validity of different kinds of tests. Through his development of these tests in his *Inquiries into Human Faculty and its Development* (1883), Galton drew public attention to the possibility of measuring intelligence.

Galton's influence spread to the United States through the work of James Cattell. It was Cattell (1890) who coined the term "mental test," and who emphasized that test administration procedures must be standardized if scores were to be used to compare people's performance over time. However, Cattell's expectation that his own battery of simple reaction tasks (involving reaction time, color vision, sensitivity to pain, etc.) might be predictive of academic success was challenged by results from studies by Sharp (1899) and Wissler (1901). Wissler's study, one of the first large-scale applications of the correlational method, showed that scores from tests that Cattell had administered several years earlier were poorly correlated with later grades of freshmen students at Barnard College. Although the designs of both Sharp's and Wissler's studies were later found to contain serious flaws, their findings raised serious questions about the efficacy of measuring human intelligence with simple tests of mental reactions.

After a long period of studying the various tasks employed by Galton and Cattell, the French psychologist Binet came to the conclusion that tests resembling the mental activities of everyday life were a more promising way to measure intelligence. As Sternberg (1985) put it, Binet countered the Galton-Cattell emphasis upon mental orthopedics with an emphasis upon higher mental processes. The *Scales of Intelligence*, which Binet introduced with Simon in 1905, consisted of a series of tasks designed to measure reasoning and judgment, capacities that were thought to transcend mere physiological responses.

For Binet and Simon, intelligence entailed three aspects: *direction*, or knowing what needs to be done and how to do it; *adaptation*, or selecting and monitoring strategies as we move toward realizing a goal; and *criticism*, or recognizing when we are performing poorly and changing our behavior to improve our performance. It is of interest that this characterization of intelligence is as relevant today as when Binet and Simon first proposed it, especially in light of current interest in metacognition and processing strategies.

As has often been recounted, the *Scales of Intelligence* had their origins in the Paris schools of the early 1900s. The French Minister of Public Education commissioned Binet to create tests that would identify the mentally defective, as well as provide a basis for instructing them. As academic success was closely associated with the ability to read and write, the tests Binet found to be the best predictors of these accomplishments contained, not unnaturally, verbal materials. Indeed, from the very beginnings of the testing movement, tests of intelligence were heavily biased in favor of verbal and numerical material.

Binet and Simon constructed their *Scales of Intelligence* by first creating a pool of diverse items obtained largely from their knowledge of teachers' expectations and their own observations of children. This pool contained such tasks as word definitions, analogies, comparisons, reasoning items, and a few tasks that required reasoning about figures (e.g., paper-folding). By administering these tasks to a large number of children of different ages, they were able to order them in a progression of difficulty that corresponded to the average performance at different ages. Through comparing a specific child's performance against these norms, one could then establish the mental age of both normal and mentally defective children.

Thus, the development of statistical techniques, the initial reaction measures devised by Galton and Cattell, the need for some way to identify ability in an increasingly industrial and technological society, and the eventual development of the *Scales of Intelligence* all combined to set the stage for what has become known as the mental testing movement. On this stage were to come the different models interpreting intelligence in general, as well as spatial intelligence specifically.

Phase 1: 1904–1938

The first phase of the measurement of spatial intelligence began approximately a year before the publication of Binet's and Simon's *Scales of Intelligence*, and lasted for more than thirty years. In 1904, Spearman published

> a strikingly original interpretation of some data that he collected in a village school in Hampshire, England. Setting up a correlation matrix among test scores and academic ranks, Spearman noticed that they could be arranged and analyzed in a special way, "hierarchically," to show that all variables measured just one "factor" in common but to different degrees (Carroll, 1982, p. 37).

From this finding, Spearman derived what he called his "Two Factor" theory, whereby

> each test of a set is regarded as measuring one "general" factor in common with all other tests and, in addition, a specific factor that is unique to the test (ibid.).

Although Spearman's theory did not attain prominence immediately, it later became a focus of international controversy as well as a starting point for rival models based upon different mathematical assumptions.

Spearman's Model

As indicated above, Spearman (1904) proposed a two-factor model of intelligence to account for patterns of correlations observed among different tests in a battery. He argued that each test could be analyzed into a general ability factor (G), which pervades all intellectual performance, and into several specific factors (S), which were common to particular tests. Spearman insisted that if group factors representing specific abilities (e.g., spatial) existed, then such factors would disturb the pattern of intercorrelations among tests so that the requirements of the Tetrad-difference equation would not be satisfied. The Tetrad-difference equation must approximate zero throughout the matrix of correlations if the hypothesis of a single common factor is to be supported. Spearman believed that (G) was the only independent factor. Although he hesitated to identify it with general intelligence, his actual applications were tantamount to that.

With respect to spatial ability, Spearman's model worked against the identification of a group factor of spatial ability over and above the general ability factor. Furthermore, it did not encourage the development of tests that might lead to evidence for the existence of such a factor.

In his later writings, Spearman interpreted the (G) factor as representing the ability to grasp relationships quickly and to use them effectively. Since his interpretation of (G) was so broad, he was able to account for the ability to solve virtually any kind of intellectual problem. For example, the ability to solve verbal reasoning problems assumed the ability to understand the relationships between the elements of such problems. The subsumption of specific abilities in (G) constituted what Spearman believed to be the law of universal unity of intellectual function (Sternberg, 1985). At the same time, because he argued that specific abilities were independent of (G) and of each other, his model was also able to account for the observed fact of individual differences over and above those attributable to (G).

During the thirty years in which this model dominated psychometrics, researchers using the Tetrad-difference method to analyze their data failed to find evidence supporting the existence of group factors. This repeated failure did not mean the rejection of the idea, however. Indeed, in the early 1920s, some evidence for a group factor was found in early studies of mechanical ability (see Kohs, 1923).

Since many technical occupations required the manipulation of concrete materials, early efforts to measure mechanical ability were undertaken with *performance tests*. Performance tests differed from the Binet and Simon *Scales of Intelligence* in that they consisted of manual dexterity tasks such as wire-bending, finger-tapping, or the manipulation of nuts and bolts. Some required the assembly of an object or the completion of a picture puzzle, and thus involved the *mental* manipulation of concrete materials. McFarlane (1925) developed a performance battery that included fitting together a number of wooden parts to make a complete object. With this task in her battery she found evidence for a spatial factor for boys but not for girls, and concluded that her tasks "measured an ability whose uniqueness lay in the fact that those persons possessing it in high degree analyze and judge better about concrete spatial situations than do other individuals who perhaps excel in dealing with more highly abstract symbols." Spearman, in commenting upon this study, suggested that her results probably reflected the fact that boys had the advantage of greater experience with construction activities.

Goddard's (1910) translation of the Binet-Simon *Scales of Intelligence* into English was a major event in the history of mental testing on both sides of the Atlantic. As Tyler (1976) notes, in the decade before World War I there was a substantial shift from qualitative to quantitative thinking about intelligence; a shift that also drew attention to the *products* of tests, rather than the *processes* required to score well on them. There was also a shift from requiring a definite answer to calling for the recognition of an answer in a multiple-choice format (Sternberg, 1982, p. 34).

With the introduction and widespread use of instruments like the *Scales of Intelligence* also came an increasing awareness of their limitations. The *Scales of Intelligence*, for example, were not satisfactory when administered to deaf subjects, to immigrants who could not understand the English language, or to those who had difficulty in expressing themselves verbally. These limitations, in part, gave rise to an increasing use of supplementary performance tests. The *Healy-Fernald* (1911), *the Woolley Scale* (1915), and the *Porteus Maze Scale* (1915) are examples of these (Whipple, 1915).

However, the administration of these non-language performance tests was not easily accepted by those in the testing movement, partly because Spearman regarded them as inferior measures of (G), and partly because of widespread bias in favor of verbal measures. This bias was clearly evidenced in 1918 when the United States Army initiated the first large-scale testing program to screen military personnel. The first test in this program, *Examination Alpha*, was developed as a personnel classification battery. It contained tests with mostly verbal content, and was not very effective in screening the large numbers of uneducated or illiterate recruits and draftees. So, a second battery of tests, *Examination Beta*, was developed to classify these men. This battery contained a variety of performance tasks and paper-and-pencil adaptations, such as Cube Analysis (block counting), Geometrical Constructions (paper formboard), and Line Tracing (paper mazes). *Examination Beta* is of historical interest, primarily because it was the first non-language battery of tests to be administered to a very large number of subjects. It was also of interest because the battery was administered to groups of subjects rather than to individuals one at a time.

As Wolman has noted, the success of the Army Testing Program was not lost on educators or industry. Beginning in 1918, group tests containing adaptations of this battery were published by a number of different authors: e.g., Dearborn's (1920) *Block Counting Test*, Stockbridge's and Trabue's (1921) *Line-Maze Test*, Thorndike's (1919) *Visual Memory Test*, and Haggerty's (1920) *Memory for Designs Test*.

> By 1925 at least fifty such group tests had been published, and many more were to follow in the succeeding decade. In 1918 Otis published his Scale; by 1923 Terman estimated that two million school children had taken the test (Wolman, 1985, p. 829).

During the 1920s there was a tremendous increase in testing and in the development of tests for persons of all ages. Unfortunately, this increase was not accompanied by an equal development in psychometric techniques, a lack that engendered skepticism among those who found that many tests failed to predict performance in schools or occupations.

The 1920s also saw considerable debate about the nature of intelligence in general, and about the relationship between non-language and language tests in particular. On the one hand, Terman (1921) among others argued that non-language or performance tests could not measure intelligence as he defined it because "an individual is intelligent in proportion as he is able to carry out abstract thinking." According to Terman, intelligence was best measured by tests that involved the use of language or other symbols. Since non-language tests contained "concrete materials," they could not be expected to tap the more abstract levels of intellectual activity.

On the other hand, Pinter and Paterson (1917), Drever and Collins (1928) and others argued that it was a mistake to think that intelligence only involved language or abstract symbols. Drever and Collins, in particular, suggested that non-language or performance tests that required complex reasoning might, in fact, be superior to a scale of verbal tests. As Wechsler (1950) put it several years later,

> the historic and continued objection to spatial tests is not that they measure too much but because they do not measure enough, or at least well enough. Thorndike said our tests measure intelligence to be sure, but that there is not just one unique but several kinds of intelligence, namely abstract (an ability to work with symbols), social (an ability to deal with people) and practical (an ability to manipulate objects).

Although during the 1920s theorists discussed the question of whether intelligence was best measured by verbal or non-verbal tests, the persistent domination of Spearman's two-factor theory reduced the significance of the issue until evidence was obtained for the existence of a group factor over and above a general ability factor.

McFarlane's demonstration of a specific factor notwithstanding, it was not until 1931 that Stephenson demonstrated statistically the existence of a verbal group factor in addition to a general ability factor. From that time onward, psychologists began to pay serious attention to the development of tests for the measurement of a range of possible group factors. Brown and Stephenson (1933), for example, were responsible for developing several of the paper-and-pencil tests that El-Koussy was to employ in his seminal dissertation research two years later in England.

El-Koussy (1935) gathered twenty-eight tests covering a wide range of abilities, and administered this battery to 162 boys between 11 and 13 years of age. The battery included both perceptual and reference tests, the latter entailing verbal inference, letter cancellation, letter series, and verbal analogies. Using a modification of Spearman's Tetrad-difference technique, he partialed out the influence of (G) from the table of correlations by means of the scores in the reference tests for (G). He concluded that there was evidence for the existence of a factor in eight of the perceptual tests but not in the others. His main conclusion was stated as follows:

> There is no evidence for a group factor running through the whole field of spatial perception. . . Spatial tests are primarily tests of (G). But some spatial tests involve a group factor over and above their (G) content. This group factor, called the (K) factor, receives a ready explanation in terms of visual imagery (p. 89).

El-Koussy concluded that this (K) factor represented "the ability to obtain and the facility to utilize visual spatial imagery." These facilities appeared to be a unique requirement for (K) tests.

El-Koussy's dissertation study is noteworthy for two reasons: it was one of the first large-scale perceptual studies by an individual in which a variety of tests were employed; and it was one of the first efforts to obtain evidence for a group factor of spatial ability. It is also interesting that he intimately associated visual imagery with his interpretation of spatial ability.

Meanwhile, in America, there had been growing dissatisfaction with both Spearman's Tetrad-difference methodology and his two-factor model of intelligence. In 1928, Kelley tested children between 10 and 16 years of age, and obtained evidence for a spatial factor. In another study he found evidence for a factor involving the "sensing and retention of geometric forms," which he distinguished from another factor described as requiring "a facility in the mental manipulation of spatial relationships." Kelley added that "we are no longer concerned with does one factor suffice but with how many factors suffice."

El-Koussy later criticized Kelley's conclusion by pointing out that "no attempt was made to determine whether such factors cover the whole realm of space perception. It is preposterous when a psychologist applies only two tests. . .and draws a conclusion for a spatial factor" (El-Koussy, 1935, p. 29).

However, growing dissatisfaction with Spearman's two-factor model was also prompting questions about the usefulness of (G) in predicting vocational success. Clark Hull (1928), for example, wrote that

> inasmuch as the very specificity of some factors must preclude their being reached by tests, the two-factor theory would preclude the possibility of forecasting by means of ordinary tests whether a person would be more effective in one aptitude than another (p. 117).

By the early 1930s, different spatial tests were developed and included in batteries that were designed for the specific purpose of examining the predictive relationship between test performance and academic or occupational competency. Brigham (1932), for example, included string length estimation and block counting tests in early versions of scholastic aptitude examinations for use in the college admission process. Similarly, the Minnesota Employment Stabilization Research Institute developed and administered a range of tests to large numbers of workers in different occupations. Standard-score norm tables were constructed for each test, which made it possible to compare an individual's scores to an average-score profile for various occupations. When Paterson et al. (1930) undertook to compare scores from mechanical ability tests with grades in junior high and technical schools, spatial tests were found to correlate .60 with success in shop courses.

Although Spearman (1927) himself came to acknowledge the existence of overlapping factors, he never abandoned his original belief in the existence of an overriding general factor of intelligence. In America, however, there was a growing willingness to consider the existence of group factors. As Carroll (1982) described it, while various mathematical techniques were advanced, chiefly in Great Britain, to show that correlation matrices often contained evidence of the influence of "group" factors in certain sets of tests. . .the methods developed to analyze correlation matrices for group factors were somewhat subjective and cumbersome (p. 69).

In America, this willingness to consider alternatives to two-factor theory found expression in the publication of Thurstone's (1931) multiple-factor methodology, and took form in the introduction of his primary mental abilities model in 1938.

Thurstone's Model

In the early 1930s, Thurstone developed methods of factor analysis that made no initial assumptions about the existence of either a general factor or group factors. Rather, these methods were intended as a means for arriving at the simplest and most meaningful description of data to which they were applied, whether the data contained only one or several factors. Spearman had concentrated upon the *correlation matrix* and found evidence, in terms of the Tetrad-difference criterion, for the existence of both (G) and for specific factors for each test when all tests in a battery were of similar cognitive content. By contrast, Thurstone concentrated upon a *factor matrix* extracted from correlational data, and found evidence for multiple factors from batteries of tests with different cognitive content. His methodology involved two steps: the extraction of factors, and the rotation of factor axes. Using his centroid method of factor extraction, he would extract one factor at a time from a correlation matrix. The first factor usually had the highest mean loadings, and each successive factor had smaller loadings as a rule. Successive factors were called *bi-polar*, because roughly half of the tests received negative signs. Thurstone's purpose in rotating axes was to achieve simplicity of factorial description such that each test would have high loading on a minimum number of factors, and zero loadings on the remaining factors.

Perhaps the reason for the acceptance of Thurstone's methods was the fact that his first major application yielded such interesting results. In 1938, he published his model of primary mental abilities, almost eight years after introducing his multiple factor analysis methods. In the interim he developed several new paper-and-pencil tests and, in 1937, administered 56 of these tests to 218 college students or graduates. From these data, he extracted 13 centroid factors from the correlation matrix, and graphically rotated 12 of them to achieve simple structure (see Guilford, 1967, for elaboration). Nine of the factors were then interpreted as representing distinct mental abilities. He labeled these primary ability factors: space, verbal comprehension, word fluency, number facility, induction, perceptual speed, deduction, rote memory, and reasoning. Thurstone described the factor labeled "space" as one that "required a facility in spatial or visual imagery." He thought that the distinguishing characteristics of tests loading upon this factor were a "facility in holding a mental image and mentally twisting, turning, or rotating it to a different position and then matching this transformed image with a suggested solution" (Thurstone, 1941, p. 21).

El-Koussy's identification of a (K) factor and Thurstone's identification of a "space" factor as a primary mental ability, together can be said to mark the end of the first phase of this historical summary. Where something like a space factor had been thought to exist for almost thirty years, and where numerous paper-and-pencil tests had been devised in hopes of measuring it, El-Koussy's and Thurstone's work finally established its existence and provided researchers with tests for measuring it. Moreover, Thurstone's criterion of "simple structure" not only served as a guide for the rotation of factor axes, it also assisted in the selection of tests to be included as markers for subsequent studies of intelligence. Indeed, the proliferation of spatial factors that occurred in subsequent years would never have been possible without Thurstone's methodology and his model of independent abilities.

Phase 2: 1938–1961

The second phase in the mental testing movement lasted for somewhat more than two decades. While El-Koussy's and Thurstone's studies in the 1930s may be viewed as an end to Phase 1, the effort to identify a spatial factor, they may also be regarded as the beginning of Phase 2, the effort to determine how spatial factors differ from one another. This second phase was also distinguished by the development of many paper-and-pencil tests and the undertaking of several large-scale research programs to specify differences between multiple space factors.

Thurstone's description of a primary mental ability factor as a space factor was similar to Kelley's earlier description inasmuch as both appeared to require the mental manipulation of spatial relationships. When Thurstone later included several additional tests in another study, he found that while spatial tests (e.g., the Flags test) loaded upon a spatial factor as expected, a number of non-spatial tests such as syllogisms and verbal classification also loaded on it. He explained this apparent contradiction by suggesting that subjects probably *visualized* items in these non-spatial tests in order to obtain solutions. This explanation has since drawn the attention of those interested in the spatiality of such tasks (Huttenlocher, 1968, and others), in differences between visual and verbal processing (Furby, 1971), and those who hold that "all that is spatial need not be figural." (Lohman, 1985).

In the late 1930s, the United States Employment Service undertook to develop measures that would aid in the screening of personnel for different occupations. During the early 1940s, alternatives to Thurstone's tests were studied, and a combination of a surface development and figure rotation task was chosen to represent the spatial factor in the General Aptitude Test Battery (GATB). The GATB has since been administered to many thousands of subjects, and its spatial subtest included in hundreds of validity studies. The GATB continues to be administered today. The strength of the battery as a screening device has been its widespread administration and careful revisions; its weakness has been the fact that a single subtest is employed to represent the spatial domain.

In 1941, Thurstone and Thurstone administered 63 tests of different cognitive materials to 710 eighth-grade children in Chicago. As expected, several primary mental abilities were identified. In interpreting their results, however, the Thurstones distinguished between spatial and perceptual processes. In particular, they noted that

> the space factor is found in tests which require that the subject manipulate an object imaginatively in two dimensions or three dimensions. It is quite different from perceptual processes which require only the perception of detail in a flat surface, and which do not require the imaginal movement in two or three dimensions. The psychological question is whether the spatial factor is somehow related to kinesthetic imagery (Thurstone & Thurstone, 1941, p. 22).

In a second phase of this study, the Thurstones selected three reference tests for each of seven of his primary mental abilities factors and administered this battery to a different group of eighth-graders. Each set of reference tests was found to load upon each of the seven factors for which they were selected. Anderson, Fruchter, Manual and Worchel (1954) subsequently criticized the Thurstones for not including different tests in their validation study.

Ever since 1938, an increasing number of researchers have employed the Thurstones' tests for primary abilities. Some, in particular, have administered them to study the effects of stimulus dimensionality of test content upon factor structures. Moffie (1940), for example, paired two-dimensional paper-and-pencil tests with similar three-dimensional performance tests (paper formboard versus wooden formboard). Although no stimulus dimensional differences were found, Moffie's study was one of the first to question whether a common ability underlay the processing of spatial information in different dimensions of test content.

With the onset of World War II, the Thurstone tests were greatly in demand by those charged with screening large numbers of military personnel. Additional spatial tests were devised, many by psychologists in the United States Army Air Force Research Program, to classify trainees who would learn to fly and maintain aircraft. Results from a series of studies in which a large number of perceptual, mechanical, and spatial tests were administered were interpreted as providing support for the existence of two, possibly three, spatial factors. The most persistent cleavage appeared to set apart one cluster of tests that measured *spatial relations* from another that measured *visualization*. With respect to the spatial relations factor, Guilford and Lacey (1947) wrote that

> it is generally agreed that the factor is a spatial one, but beyond this point there are divergences of opinion. Two hypotheses which have been proposed deserve consideration here. According to one, it is an ability to make discriminations as to the direction of motion. The term "discrimination" here does not carry the usual connotations of perceiving small differences, for obviously the spatial distinctions called for in the test are very gross. The decisions are frequently merely between up and down, left and right, and in and out.
>
> Another hypothesis is that the ability is concerned with the general apprehension of spatial relations. Either stimuli or responses or both in the spatial tests are arranged in spatial

patterns, and there is frequently a systematic relationship between order in the stimulus and order in the response. In such a test, therefore, the essence of the spatial factor could be 1) ability to perceive visual-spatial arrangements, 2) ability to organize movements in spatially-determined order, or 3) ability to relate specific locus or arrangement within the response pattern. The second and third characteristics would apply to the psychomotor tests but not to all printed tests (pp. 478–479).

By contrast, there was general agreement about the characteristics of the visualization factor. It was described as follows:

All of the tests heavily saturated with the factor seem to involve a visual manipulative ability. In solving the problems, it is necessary to move, turn, twist or rotate an object or objects and to recognize a new appearance or position after the prescribed manipulation has been performed (ibid., pp. 291–292).

According to Anderson et al. (1954), a third possible spatial factor was characterized in terms of a further breaking down of the visualization factor. Specifically, it was identified as a factor restricted to some psychomotor tests and to a few paper-and-pencil tests such as Thurstone's Flags or Hands tests. This factor appeared to require an appreciation of left-hand, right-hand discrimination. But as Zimmerman noted, "there seemed to be no appropriate terminology left to describe the remaining factor" (Zimmerman, 1954, p. 396).

A side-effect of these Army Air Force studies was the fact that a large number of psychologists were brought together to study spatial ability for screening and classification purposes. After the war, several of them continued to re-analyze or to continue to collect spatial data. Roff (1951), for example, reported results from the administration of 44 experimental aptitude tests to a population of pre-aviation cadets. Fifteen factors were extracted by the multiple-group method. Roff found four factors in a space category, and labeled them as *orientation*, *visualization*, *space*, and *visual pursuit* factors. What he called orientation was similar to the spatial relations factor found in earlier Army Air Force studies; the space factor was similar to Thurstone's space factor.

In other studies, while Fruchter (1949) found evidence for a visualization factor in a postwar study of aviator populations, he did not find evidence for a spatial relations factor. By contrast, Michael Zimmerman and Guilford (1950), comparing data from two Army Air Force populations, found evidence for a spatial relations factor but not for a visualization factor. Despite these apparent contradictions, Guilford and Zimmerman (1948) decided to include both a spatial relations and a visualization subtest in their aptitude test battery designed for general guidance and vocational placement.

In 1949, Thurstone and Thurstone reported results from a study conducted for the United States Navy, which was designed to tease apart components in the complex known as *mechanical ability*. Among several hypotheses tested, they were especially interested in comparing performance on tasks requiring the ability to visualize stationary solid objects, with those requiring the ability to visualize solid objects moving through three-dimensional space. They administered 32 group tests and 25 individual tests to about 350 high school boys, and found evidence for three spatial factors. Space factor *S1*, with appreciable loadings on such tests as Cards and Figures, was described as representing "an ability to visualize a rigid configuration when it is moved into different positions." Factor *S2*, with moderate loadings on such tests as Paper Puzzles and Surface Development, was described as representing "an ability to visualize a configuration in which there has been movement or displacement among the parts of configuration." Factor *S3*, described as operating with problems involving left-right identification, appeared similar to the Army Air Force factor of spatial relations. The difference between S1 and S2 was attributed to the static or kinetic imagery involved in test solutions.

French (1951) looked at many factorial studies that had a number of tests in common in order to identify common factors. He concluded that evidence existed to support three separate spatial factors. The first of these, *spatial*, he saw as an "ability to perceive spatial patterns accurately and to compare them with each other." He considered the second factor, *orientation*, in need of further clarification, but felt it represented the ability "to remain unconfused by the varying orientations in which a spatial pattern may be presented." The third factor, *visualization*, he interpreted as representing an ability "to comprehend imaginary movements in three-dimensional space or to manipulate objects in imagination." Although French's three-fold description was an important effort at synthesizing a fragmented subject, in subsequent studies by other researchers stray factors kept escaping it.

The proliferation of tests and factors in the 1950s caused serious research problems, one of which was the necessity of administering very large test batteries in order to accommodate so many factors and subfactors. When spatial factors were studied in more modest efforts, it was not always clear from the results whether differences existed because the abilities themselves were distinct; because the motivation of subjects toward different spatial tasks differed; because the quality of the stimulus presentation or the verbal instructions varied; or because the testing conditions differed for different studies. As Thorndike (1931) and others have repeatedly pointed out, even minor variations in test administration procedures may be as important to results as differences in subject populations or in the kinds of data entered into the factor analysis.

Although in England there were no large-scale research efforts during World War II like the U.S. Army Air Force Research program, large numbers of military personnel were nonetheless tested for classification purposes. Following the war, British psychologists struggled not so much with the existence of different spatial factors, as with the problem of devising ways to allocate children at age 11 to either grammar or technical schools (the Eleven-Plus exams). In response to this latter problem, new tests of intelligence were developed, and new models advanced to account for individual differences. Among the new test batteries were the Moray House and the NFER series; among the theoretical models were those proposed by Burt (1949) and Vernon (1950).

In 1949, Burt published a review of the evidence regarding the contribution of practical ability to trade or technical school success. He concluded that practical ability appeared at about age 11, increased in importance until about age 15, and thereafter became more specialized, possibly because of occupational interests and habits. Burt also concluded that practical ability included both a mechanical and a spatial factor. With regards to the latter, he wrote:

> The spatial subfactor appeared to be concerned essentially with the ability to perceive, interpret, or mentally rearrange objects as spatially related. Some of our test results, obtained with boys at trade schools and with older engineering apprentices, suggests the need for distinguishing between two factors of this kind, namely, a factor for static spatial relations and a factor for kinetic spatial relations. Kinetic relations are apprehended mainly by aid of movement of the observer's own organ, especially his eyes. But even when spatial relations are apprehended statically, their correct perception rests upon schemes built up kinesthetically as a result of movement of the observer's own organ. As regards static relations, there seems to be a further distinction between the ability to perceive and visualize relations in 1) two-dimensions only, and 2) in three-dimensions (pp. 188–189).

Like Thurstone and Thurstone (1941), Burt clearly was trying to account for differences in spatial aptitude in terms of stimulus dimensionality and static or dynamic aspects of visual imagery.

In addition to addressing the Eleven-Plus question of school allocations, Burt also proposed a model of intelligence based upon results of research with military personnel and others.

Burt's Model

Burt's model of intelligence was conceived as a hierarchy based upon levels of successive dichotomies, with each subdivision of a higher factor yielding two lower factors. At the top was a (G) factor, beneath which, at a *relations* level, were both (G) and a practical factor. Both factors at the relation level were in turn split into four factors at an *association* level. Below these were eight factors at a *perception* level, and below the perception level were sixteen factors at a *sensation* level. A graphic representation of this model is presented in Figure 3-1.

Burt's model is interesting because it was an attempt to account for group factors within a modification of Spearman's earlier scheme. It was also interesting because he developed factor analytic procedures that enabled researchers to extract first a (G) factor and then group factors of successively smaller breadth. These techniques were influential with respect to the model advanced by Vernon a year or so later.

Vernon's Model

In 1947, Vernon reported the results of a study in which 1,000 military conscripts were administered a battery of thirteen tests, and found that a (G) factor covered more than twice as much variance as all group factors combined. Moreover, after removal of (G), tests seemed to fall into two main categories; a verbal-numerical-educational category (which Vernon called the v:ed factor), and the practical-mechanical-spatial physical category (which he called the k:m factor). Thus,

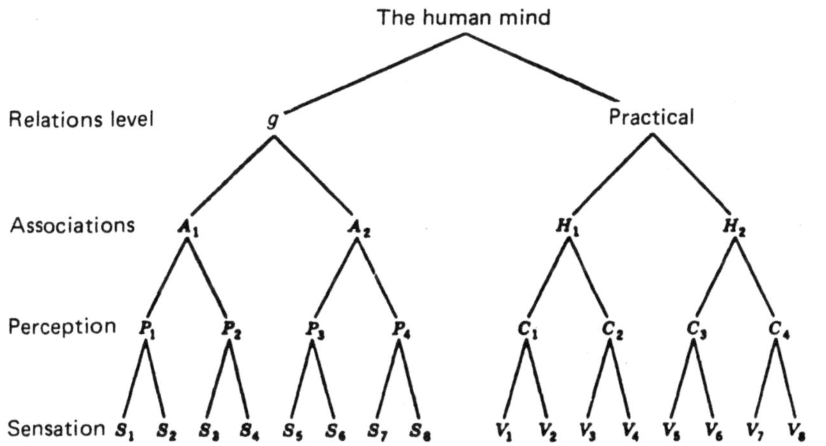

FIGURE 3.1. Burt's hierarchical model of ability factors. (Reproduced from Jensen, A.R. [1970]. Hierarchical theories of mental ability. In: B. Dockrell [Ed.]. *On intelligence*. New York, Methuen, p. 131.)

Vernon found a first approximation of mental structures in a hierarchy which resembled a genealogical tree (Vernon, 1950, p. 23). He felt that the k:m factor was an aggregate of group factors that were heterogeneous and less influenced by the effects of training or schooling. Among children or dull adults, spatial and numerical tests appeared opposed, apart from (G), but for normal adults both spatial and non-verbal tests of (G) tend to link up with mathematical ability. Thus, the numerical factor that seems to be in the verbal-educational cluster at low levels shifts over to the non-verbal cluster at higher levels. As for spatial ability, Vernon insisted that there was only one underlying factor, and that the American plethora of spatial factors was more confusing than helpful (Vernon, 1961, p. 75). A graphic representation of Vernon's model is presented in Figure 3.2.

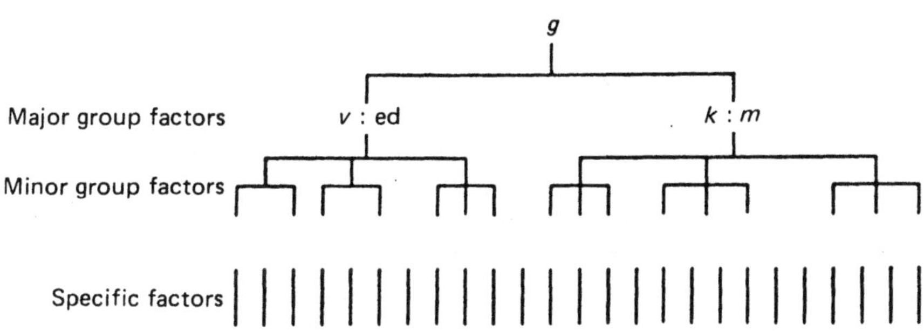

FIGURE 3.2. Vernon's hierarchical model of ability factors. g = general factor; v:ed = verbal-educational aptitude; k:m = spatial-mechanical aptitude. (Reproduced with permission from Jensen, A.R. [1970]. Hierarchical theories of mental ability. In: B. Dockrell [Ed.]. *On intelligence*. New York, Methuen, p. 131.)

An advantage of hierarchical models like Burt's and Vernon's is that they contain more information than primary factor models such as Thurstone's, since in addition to primary factors, they contain the broader, more general factors. Moreover, as Snow (1978; 1980) has pointed out, these broader factors are often more amenable to analysis and understanding than are primary factors. A disadvantage of hierarchical models is that

> it is quite difficult to use the model in such a way that individual scores on factors at different levels are estimated and used as predictors; instead the model has most commonly been used for selecting and classifying tests (Gustafsson et al. 1981, p. 17).

In the thirty years since the publication of Vernon's model, there have been many studies that have pitted verbal or academic factors against spatial or practical ones. Although Vernon himself warned against such dichotomizing of abilities, many researchers continue to use factor analytic methodology, which results in a bi-polar division of group factors. In 1979, for example, McCallum, Smith, and Eliot administered a battery of fifteen tests to secondary school students in England. Not only was evidence for a spatial factor reported, but it was also found to be statistically unrelated to a factor of verbal comprehension. When such results are published, it should not be surprising that other researchers assume that spatial or verbal tasks may be representative of dichotomous domains of behavior.

In America during the 1950s, several important studies were undertaken to examine apparent differences between factors of spatial ability. Some of these studies have already been cited (e.g., Roff, 1951; Michael et al., 1950; and Fruchter, 1949), but three others deserve mention.

One is Zimmerman's re-analysis of Thurstone's (1938) data, and his investigation of task complexity. In 1953, Zimmerman took Thurstone's data and further rotated its centroid axes toward simple structure. He found evidence for two spatial factors, instead of the one originally reported by Thurstone. The first factor, defined by such tests as Flags or Cards, appeared to be similar to Thurstone's original space factor. The second, defined by tests such as surface development and paper-folding, he labeled *visualization*. The difference between the two factors seemed to be that tests of spatial relations were timed and less complicated, while tests of visualization were not timed and were more difficult.

In Zimmerman's (1954) study of the effect of task complexity, he demonstrated that by increasing the difficulty of items on an Army Air Force experimental test, the same type of item could be made to load in succession upon a perceptual speed factor, a space factor, and a visualization factor. In the simplest form of the Visualization of Maneuvers Test (1947), subjects selected from five alternatives the picture that correctly depicted the position of an airplane after it had completed a given maneuver. In the moderately difficult form, they had to visualize two maneuvers to reach a solution, and in the most difficult form, three maneuvers in succession. Zimmerman concluded that each of the three forms of the test required somewhat different response procedures. Moreover, he concluded that he had found evidence for a continuum of task complexity, and that the ability to *visualize* relationships required more intellectual effort than carrying out the same task at a spatial orientation or perceptual speed level of complexity. He proposed that his results indicated that the Thurstones' S1 and S2 factors could be interpreted in terms of differences on this continuum of task complexity.

Another study was undertaken by E. S. Barratt (1953), who recorded verbal retrospective accounts for a variety of spatial tests, and found that subjects reported different strategies for solving them. Specifically, he administered seven spatial and three verbal tests to 84 college males. Four centroid factors were extracted from the correlational matrix, and then rotated to orthogonal simple structure. Barratt found that the three verbal tests defined the first factor; Thurstone's Cards, Flags, and Figures tests defined a second; a Chair-window test and the Industrial Aptitude Spatial Orientation test defined the third; and the DAT Space Relations and Guilford-Zimmerman Spatial Orientation test defines a fourth. He labeled these as *verbal, spatial relations, spatial orientation*, and *visualization* factors, respectively. Of greater interest were the retrospective verbal reports of how each subject solved the different spatial problems. The analysis of interview protocols took several forms.

Barratt defined the spatial relations factor as "the ability to turn or rotate a given figure or part of a figure in one plane (or about an imaginary axis) to see if it corresponds to another figure in the same plane" (Barratt, 1953, p. 21). The spatial orientation factor was defined as the "ability to determine from where you are looking at an object; i.e., where one is spatially located in relationship to a particular object" (ibid., p. 22). And the visualization factor was defined as the "ability to see or observe the spatial relationships of objects involved in dynamic situations, spatial relationships in which the subject has to imagine that the object or objects involved changed their positions relative to one another" (ibid. p. 21). With respect to strategies, 82 of the 84 subjects used the method described for the spatial relations tests; 58 used the method described for the spatial orientation tests, and between 76 and 83 used the method described for the visualization tests. As Lohman noted, there was a tendency to "shift from a direct mental manipulation strategy to a more 'analytic' strategy using particular stimulus features and logical inference as item difficulty increased" (Lohman, 1979, p. 138).

Barratt's study is noteworthy because it was one of the first to go beyond factor analysis and examine the strategies used to solve different spatial problems. The fact that a problem may be solved by several distinct strategies may explain why a test will sometimes load upon one factor and sometimes on another. Finally, his results also indicate that a single subject may use different strategies to solve different items on the same test.

A third important study was undertaken by Myers (1957; 1958). Like Barratt, Myers asked college students to discuss how they solved spatial tests, but he gave fewer tests to fewer subjects. He found three sources of difficulty with respect to a Hidden Blocks and a Surface Development test: namely, a number of students failed to understand the directions, many failed to "read" the line drawings of test items, and some used imagery to solve easier problems but shifted to more "analytic" methods when the tasks became more difficult.

As several researchers have observed, the retrospective data collected by Barratt and Myers must be treated with caution. As Bloom and Broder wrote earlier with respect to retrospective reports:

It is very difficult for a person to remember all the steps in his thought processes and report them in the way they originally occurred. There is a tendency on the part of the narrator to edit the report, to set forth the process in a nicely logical order. Things seem to tie together so nicely after the problem has been solved. The narrator will usually omit errors and "dead ends" in his thinking processes. He will not remember the queer quirks and unusual circumstances which surround his thinking. Such reports generally present a coherent and well-ordered train of thought rather than the incoherent and jumbled process which may have occurred. These retrospective accounts are useful, but it must be recognized that they are rebuilt outlines of thought processes and tend to reveal only the high spots and finished products rather than the raw materials and details in a fantastically complex series of thought steps (Bloom & Broder, 1950, p. 6).

Newman (1951) makes a related point when he talks about the contribution of spatial experience to mathematical problem solving. As he pointed out, while diagrams, graphs, and other visual aids frequently guide mathematicians as they grope for a proof, once a theorem is proved, they tend to remove all traces of these aids. When the successful student then reads the proof, he reconstructs the spatial information in order to grapple with the argument. But when the unsuccessful student, not understanding the significance of this reconstruction, vainly tries to imagine "any triangle" rather than sketching a specific triangle, he usually fails to reach the proof. Thus, not only are processing differences difficult to obtain from retrospective reports, they may also be lost altogether in the solution of non-spatial problems.

In Smith's (1964) opinion, the second phase of the mental testing movement ended in 1957, not 1961, with the publication of a study by Michael, Guilford, Fruchter, and Zimmerman. These researchers reviewed the literature on spatial intelligence and made still another attempt at reconciling differences between factor interpretations. They found three grouping of factors that appeared to correspond to the descriptions from the Army Air Force research program (Guilford & Lacey, 1947), the Thurstone studies (1938; 1941; 1944; 1951), and the descriptions obtained by French (1951). The first grouping, a combination of the spatial relations-orientation factors, was thought to require the ability to comprehend the nature of the arrangement of elements within a stimulus pattern. A second factor, visualization, was thought to require the mental manipulation of visual objects in a specified sequence. And a third factor, kinesthetic imagery, was thought to involve left-right discrimination.

As Smith observed,

a remarkable feature of the American research on spatial ability is the difficulty the American psychologists are finding in clarifying the distinctions between the different spatial factors and especially between spatial relations-orientation and visualization factors. In spite of numerous studies carried out with large samples of testees and the many discussions, seminars, and symposia conducted on the problem since the early investigations by Kelley and Thurstone, it cannot be said that clear-cut distinctions have been established. Michael et al. (1957) have considered in great detail psychological aspects such as complexity of stimulus, amount of manipulation involved, the examinee's bodily orientation, the movement of parts of the stimulus object versus the movement of the entire object, as well as the relative importance of speed and power. They claim there are differences between the three factors in each of these respects, but none of the differences seems to be really water-tight (Smith, 1964, p. 95).

According to Smith (1983), the Michael et al. (1957) study represented a water mark in the psychometric tide of research on spatial factors. Although we will here define the second phase of the mental test movement as extending until Werdelin's (1961) study, Smith thought the Michael et al. (1957) study to be a more useful demarcation point because he thought that it was an excellent example of the proliferation of spatial intelligence into a host of factors, and because it was one of the last attempts to reconcile factor descriptions without subjecting earlier data to a metaanalysis or common factorial methodology.

During this second phase, Thurstone's space factor had been broken down to subfactors by researchers concerned with the classification of large numbers of military personnel. Not only were new methods of factor analysis developed, but many new tests as well. After World War II, there were several efforts to achieve some agreement about the descriptions of different factors, to examine the relation of spatial intelligence to educational success, and to move beyond factor analysis to study solution strategies from retrospective reports. Unfortunately these descriptions of spatial factors were frustrating to other researchers because they appeared at best vague and at worst self-contradictory (Clements, 1983), and this frustration caused many researchers to turn away from the study of spatial intelligence. As will be seen in the description of the third phase, attention shifted from efforts to simply define different spatial factors, to efforts to account for sources of variation in test performance.

Phase 3: 1961–Present

Within the last twenty-five years, interest has shifted from establishing the existence of spatial factors and then distinguishing among them, to ascertaining how they relate to other abilities. With the exception of Guilford's research, this phase has been marked by a decline in large-scale efforts to define or interpret spatial factors, and by an increase in the amount of research undertaken to discover sources of variance in performance on spatial tests. Several of these latter studies have emerged as research topics in their own right, with their own models, preferred methodologies and tasks, and cumulative literatures. The controversy about sex-related differences in spatial ability, for example, has produced a number of books and a large number of articles since Maccoby's and Jacklyn's (1974) review.

Major events during this most recent phase have included the popularization of Piaget's development account, the rise of cognitive psychology, the development of the computer both as an analogue of mind and as a means of rapid processing of data, and most recently the re-analyzing of data, using a common statistical methodology, from older factor analytic studies.

Werdelin's (1961) study of geometrical and spatial factors is a good starting point for this third phase, both because it employed factor analysis and because it looked at sex-related differences. Werdelin administered 23 tests to three samples totaling 453 high school students. The data from each sample were analyzed separately using Thurstone's centroid method, and six factors were extracted in each analysis. For the boys, he found spatial factors that he interpreted as being similar to the Thurstones' (1949) Sl and S2 factors. He described the first spatial factor as representing "the ability to see a whole configuration or Gestalt" (Werdelin, 1961, p. 77); and the second as requiring "the breaking down of the configuration into elements, and then the working with one element at a time" (ibid., p. 79).

For the sample of girls, only one spatial factor was found. When he inspected test and item content, he noted that girls differed from boys in their ability to organize and re-organize visual structures, but not in their ability to mentally manipulate these structures.

This study is interesting for several reasons. First, like Thurstone, Werdelin included a variety of tests with different cognitive content in his battery and found that tests of number series and verbal syllogisms loaded upon a spatial factor. Second, when commenting upon Zimmerman's complexity hypothesis, he noted that spatial ability was negatively related to tasks requiring reasoning procedures that rapidly become automated, as in the solution of simple equations or mechanical computations. Third, like Barratt and Myers, Werdelin made some interesting observations about the relationship between spatial task demands and subjects' solution processes. In particular, he wrote that

> when the subject has to solve a visual problem, there are always two starting points: the visual material, which has to be worked on, and the verbal instruction, which tells the subject what to do and initiates the performance. . . . The subject must comprehend the visual organization of the material; he must perceive the whole figure, find the relation between the different elements of the structure, he must form a gestalt out of often roughly-drawn lines and figures, from the disparate elements a figure may consist of, a gestalt that will have its meaning from the problem involved in the tasks. Then he must often re-organize the given structure, if it does not suit the problem in question. Finally, he must often "manipulate" the spatial figures if this is the task of the test in question. These mental processes are involved in all visual tests, and these processes are not isolated from each other but form integrated parts of a united mental process. But in different tests different parts of this process are stressed (Werdelin, 1961, p. 77).

Lastly, Werdelin's reporting of sex-related patterns of factor loadings anticipated a more recent controversy, as will be seen.

In 1964, the publication of a book by Smith provided scholars with a comprehensive review of the history of the measurement of spatial ability up to then. Smith's book is interesting because it was written from the perspective of the Spearman tradition; because it contains a chapter about the relationship between spatial ability and mathematics; and because it comments upon the mental processes thought to be required in the solution of spatial tests.

With respect to solution processes, Smith refers to Spearman and Jones (1950) to the effect that

> the senior author happened to notice that spatial tests can readily be performed in two distinct manners. One may be called "analytic" in the sense that attention wanders from one element of the figures to another. The other mode of operation is comparatively synthetic in that figures (or their constituents) are mentally grasped in much larger units (sometimes called wholes) (Smith, 1964, p. 203).

With this distinction in mind, Smith goes further and asserts that analytic processing, or "the explicit trial and error, checking and cross-checking of relationships between the different parts of a figure" was more commonly associated with a perceptual factor, while synthetic processing, or "the ability to perceive or to recognize the structure or form of a shape or pattern as a whole or gestalten" was more characteristic of a spatial factor. Moreover, he argues that the long-standing view of psychologists such as McFarlane (1925) and Spearman (1950) of a dichotomy between concrete and abstract intelligence has been misleading. Smith insists that if spatial ability entails the mental manipulation of whole figures or shapes, then surely it requires abstract thinking.

Both Werdelin's and Smith's observations about the solution processes required by spatial tests appeared at a time when psychologists generally were becoming more interested in perception, language, imagery, memory, and thinking or cognitive processes. Chomsky's (1957) ideas about language, Piaget's (1956) ideas about spatial conceptualization, Bruner's (1960) ideas about education, and Newell, Shaw, and Simon's (1958) ideas about the computer-mind analogy all caught the attention of psychologists and helped to shift the focus of interest from the study of behavior qua behavior to the study of the mind. Not surprisingly, many psychologists looked to various tests developed by psychometricians to measure different cognitive processes.

Toward the end of the 1960s two models of the structure of mental abilities were proposed, each derived from different factor analytic methodologies, and each supported by empirical research. Guilford's Structure of Intellect model had its roots in the Army Air Force Research Program of World War II, and evolved as the Aptitudes Project at the University of Southern California. Horn's and Cattell's hierarchical model, by contrast, had its beginnings in Spearman's laboratory during the 1930s, and subsequently evolved over the years at the University of Illinois. Both models incorporated different facets of expressions of spatial ability in their respective structures or organizations of ability. We shall look first at Guilford's Structure of Intellect model.

Guilford's Model

As mentioned, Guilford gained considerable experience with tests from his involvement with the Army Air Force Research Program during World War II. His model of ability organization was based upon a factor analytic methodology that Thurstone had employed in his own early research. Specifically, when Thurstone rotated factor axes to approximate the criterion of simple structure, factor score estimates correlated only slightly. But when Thurstone and Thurstone investigated the existence of primary mental abilities in children, they found the tests and also the factors to be intercorrelated. The Thurstones' finding led them to adopt an oblique factor model, in which correlations were allowed among the primary mental abilities. Although oblique axes may be re-analyzed to demonstrate higher-order factors, the Thurstones paid little attention to a general factor in their studies of the late 1930s and 1940s.

Guilford, preferring Thurstone's earlier orthogonal rotation of factor axes, rejected the idea of a general ability factor, and instead mounted a sustained effort to find order among the confusion of the partially independent factors that had been identified in the Army Air Force Research Program (Guilford & Lacey, 1947). As more and more factors were identified statistically, he saw that they could be related to one another in a three-dimensional classification comparable to the periodic table of elements in chemistry. Specifically, he argued that intellectual activities could best be understood in terms of the types of mental *operations* performed, the *content* involved, and the resulting *product*. Spatial ability was no longer conceived of as an independent family of related factors, but instead was to be measured by tests that possessed different combinations of operations, content, and products. Guilford made no assumptions about the relationship among abilities that shared common positions on one or two dimensions.

A graphic representation of Guilford's Structure of Intellect model is presented in Figure 3.3. Guilford accounted for intelligence in terms of four kinds of content or material, five kinds of mental operations or activities performed on the content, and six kinds of products that result from performing operations on content.

According to Guilford and Hoepfner (1971), the kinds of content or material included: 1) figural (size, color, form), 2) symbolic (numbers, letters, etc.), 3) semantic (verbal meaning or ideas), and 4) behavioral (social contexts). The operations performed on these four kinds of content included: 1) *cognitive* (recognition or discovery), 2) *memory* or recall of what is known, 3) *divergent* thinking, or looking for alternative solutions to problems, 4) *convergent* thinking, or seeking strategies that lead to correct solutions, and 5) *evaluation*, or judging the adequacy or correctness of a response.

These five operations performed upon the four types of content resulted in six kinds of products. Guilford (1959) described these products in the following terms:

> The product may be a unit of information, which is defined as a relatively segregated or circumscribed portion of information, with a "thing" character. The product may be a class, which we define as an aggregate of units of information, grouped because of their common properties. A relation, a third kind of product, is defined as a connection between units of information, based upon variables that apply to them both. A system is defined as an organized or structured composite of units of information, having interacting or inter-relating parts. The concept of "group" in mathematics would be an example in the symbolic category, as a set would be an example of a class. A transformation is some kind of change of existing or known information; a reinterpretation. An implication is some

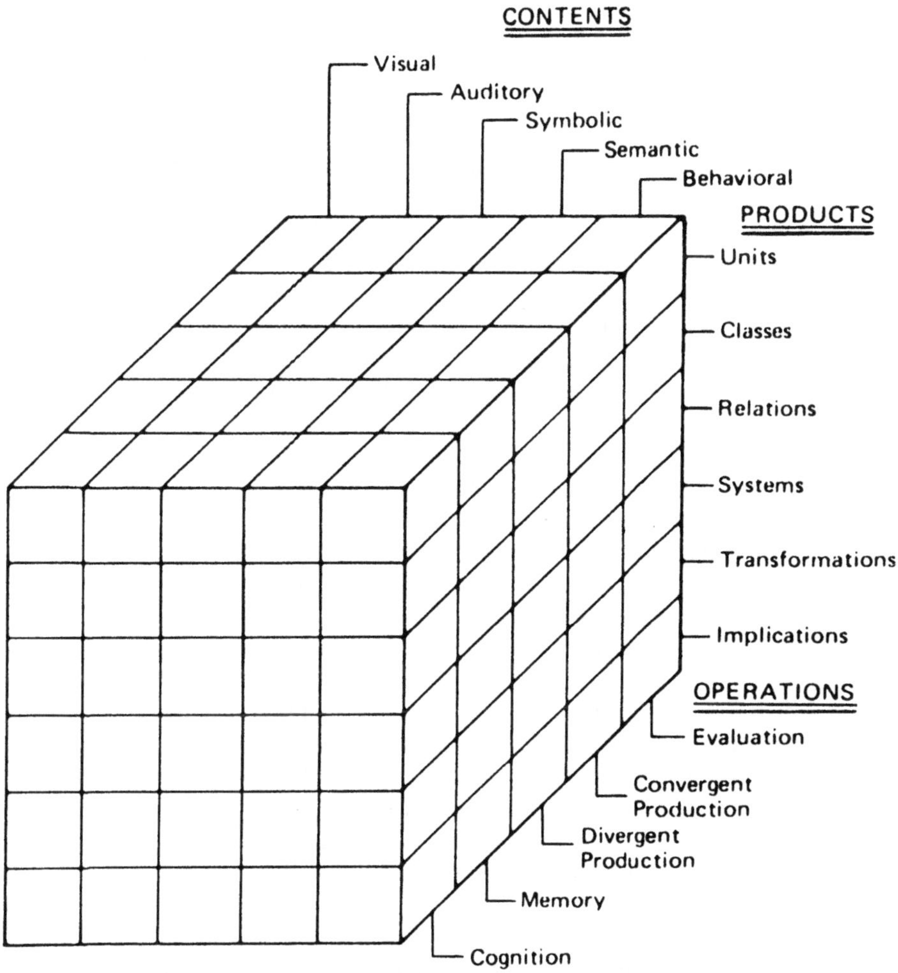

FIGURE 3.3. Guilford's structure of intellect model. (Reproduced with permission from Guilford, J.P. [1985]. Structure of intellect model. In: B. Wolman [Ed.]. *Handbook of intelligence*. New York, Wiley, p. 230.)

kind of extrapolation of knowledge. In the area of cognition, this would include such things as expectation and knowledge and antecedents or consequents. In the area of productive thinking it would include drawing inferences or conclusions.

With each ability redefined as a combination of content, operation, and product, Guilford's early model contained 120 separate abilities or factors, each of which is defined by its unique position or cell in his classification.

Guilford and his associates then set about the task of developing an astonishing number of instruments to measure the abilities in each cell of this classification. These new tests (and other putative measures) were then employed in several studies intended to determine whether they would load upon the hypothesized factors. In 1968, for example, Hoffman, Guilford, Hoepfner, and Doherty administered 72 tests representing 23 of the hypothesized factors to 250 architectural students. The resulting correlational matrix was factored by the principal factoring method, and the principal axes were then orthogonally rotated by an analytic procrustean procedure developed by Cliff (1966). The initial target matrix was formed by inserting a loading for each test equal to the square root of its commonality on the one factor it was hypothesized to measure. New target matrices were constructed after each of seven iterations of this procedure, and the 22 hypothesized factors extracted were interpreted by Guilford as support for his model.

The model has been attacked in a number of ways. Guilford has been criticized for deriving factors based upon targeted rotations in which he guided the factor solution by hypothesizing an "ideal" set of loadings for each of the tests in a sample. In this sense, he used factor analysis to confirm the existence of a hypothetical classification, rather than to discover a structural model of intellectual abilities. He has also been criticized for subjectivity which, in this instance, refers to "over-factoring to assure as many factors as desired" and using targeted rotation to assure maximum correspondence between results and the desired factor pattern (Eysenck, 1973; Horn & Knapp, 1973; 1974). Some have accused him of stretching the meaning of intellectual ability to include abilities involving behavioral content, a domain heretofore not associated with intelligence.

Guilford and his associates have been further criticized for being so logical as to be unrealistic; for their failure to account for the fact that some tests correlated with many others while some correlated with only one or two; for concentrating upon the processes believed to be required for test solution rather than for solving everyday problems; and for further splintering the construct of intelligence by dramatically increasing the number of known "primary" abilities.

On the other hand, Guilford's model has contributed to our understanding of intelligence by focusing attention upon the processes underlying intellectual activity; by encouraging the development of many new instruments for measuring different facets of intelligence; and by serving as a guide for a considerable amount of research activity. Perhaps most importantly, it underscored the temptation of psychometricians to reify more traditionally understood factors and to treat them as independent of one another.

With respect to spatial ability, the model required psychometricians to think about tasks in terms of their content, operations, and products rather than in terms of their loading upon specific factors like spatial orientation or visualization. A visualization task like the paper-folding test, for example, was reclassified in Guilford's model as a *cognitive, figural*, and *transformational* (CFT) ability. By contrast, a hidden figures test, occasionally treated as a measure of visualization, appeared as a *convergent production of figural transformation*. With this model, Thurstone's (1938) space factor clearly became the sum of many different facets.

In 1982 (and more specifically in 1985), Guilford changed his classification by dividing his figural content dimension into an auditory and a visual content dimension (and thus increasing the number of cells in his model to 150), and modified his stance regarding the rotation of factors to admit to the possibility of higher-order factors (Guilford, 1985, p. 235).

The domain of spatial ability also all but disappeared in Horn's and Cattell's (1966) hierarchical model, but for entirely different reasons. The reasons should become clear as this model is described.

The Horn and Cattell Model

As mentioned earlier, Horn's and Cattell's (1966) model had its beginnings in Spearman's laboratory before World War II. In 1943, Cattell posited the existence of two correlated general intelligences: *fluid ability* (gf), and *crystallized ability* (gc). However, for various reasons he neither elaborated upon these ideas nor subjected them to empirical tests. Later formulations of the gf-gc model, especially Cattell's (1971) triadic theory, were to depend for empirical support upon a study by Horn and Cattell (1966) which sampled a sufficiently large number and variety of first-order factors to permit meaningful second-order analyses.

Whereas both Guilford and Cattell used Thurstone's centroid method to extract factors from correlation matrices, Guilford chose to rotate factor axes to an orthogonal frame with zero correlations between factors. Cattell, by contrast allowed factor axes to be oblique (correlated) in order to achieve the best approximation of simple structure. From an oblique factor analysis, a large number of primary or first-order factors are often obtained. If the correlations between these first-order factors are subjected to another factor analysis, the matrix yields second-order factors and, if these in turn are made oblique, the process can be made to yield third-order factors. Cattell's methodology produced just such a hierarchical organization of abilities, with factors from specific tests becoming more pervasive and more general at each successive higher-order analysis.

Empirical support for Horn's and Cattell's model was obtained from a study in which a battery of tests, representing 23 primary abilities, was administered to 297 volunteer subjects. Horn and Cattell derived five second-order factors from their analysis, including fluid ability (gf), crystallized ability (gc), power of visualization (gv), retrieval capacity (gr), and cognitive speed (gc). They felt (gf) represented the basic biological capacity of the individual; (gc) represented the effects of education and training; and visualization (gv) entered into tasks with figural content, especially through such primaries as spatial visualization, spatial orientation, flexibility of closure, speed of closure, and figural adaptive flexibility. They described general visualization as being involved "in visualizing the movements and transformations of spatial patterns, maintaining orientation with respect to objects in space, unifying disparate elements, and locating a given configuration in a visual field" (Horn & Cattell, 1966, p. 254). Unfortunately, when Lohman (1979) re-analyzed their data using a nonmetric multidimensional scaling and hierarchical clustering analysis, he failed to bring out (gf) and (gv) as separate factors.

The Horn and Cattell hierarchical model differs from Vernon's earlier model in significant ways. Obviously it differs with respect to descriptions of different levels. But, as Kail and Pellegrino (1985) point out, it also differs in that Cattell (1971) later theorized about causal relationships among different ability factors and their emergence with increasing age. Perhaps most importantly, the two models differ in that Horn and Cattell based theirs on a bottom-up approach to factor analysis (from primary to broad general factors), while Vernon used a top-down approach (from general ability to group factors).

Other models of ability organization were proposed in the early 1970s in addition to Guilford's Structure of Intellect model and Horn's and Cattell's hierarchical model. Royce (1973), for example, also defined spatial ability in terms of a higher-order structure of abilities. He concluded that while data from tests of spatial visualization, flexibility of closure, spatial orientation, speed of closure, adaptive flexibility, and spatial scanning combined as one second-order factor, data from tests of spatial scanning, speed of closure, and perceptual speed combined as a different second-order factor. Royce thought these two second-order factors then combined, in turn, with a reflexibility-impulsivity factor in a third-order factor of field-articulation. The resulting hierarchy thus included both ability and cognitive style in an alternative model of intelligence. Unlike Horn and Cattell, however, Royce did not elaborate further on his structural model.

Another model evolved from French's (1951) summary of factorial investigations. In 1954 and again in 1963, French and his associates at the Educational Testing Service compiled in two successive "kits" a sampling of tests representing confirmed factors of mental ability. According to the manual accompanying the third version of the kits (1976), a factor was included if its underlying construct was found in at least three factor analyses performed in different laboratories by different investigators. The three kits were thus presented as collections of factorially similar tests, not as a model per se. Nevertheless, the kits, as they contained factors "identified with sufficient certainty to receive a name" (French, 1954, p. 200), have subsequently served the function of a model as a guide for test selection in a large number of studies.

The kits contained a variety of spatial tests for different factors. Some of these tests originated with Thurstone, others with Guilford, and still others with the Army Air Force Research Program. In the 1963 *Kit of Reference Tests for Cognitive Factors*, French, Ekstrom and Price described tests for the spatial orientation factor as requiring "an ability to perceive spatial patterns or to maintain orientation with respect to objects in space." Tests for a visualization factor were described as requiring "an ability to manipulate or to transform the image of spatial patterns into other visual arrangements." By contrast, more than a decade later, Ekstrom, French, Harman, and Derman (1976), in the third version of the *Kits*, described spatial orientation as requiring "only the mental rotation of a configuration," and tests of the visualization factor as "requiring that a figure be mentally restructured into components for manipulations." Such changes in factor interpretation over time underscores the fact that the factors themselves are dependent upon specific tests, not necessarily activities in "real life." Horn (1972) speaks to this point when he writes:

It is not always clear what we mean by a factor and what we mean by a test. We speak of a factor of visualization, and we know it is indicated by tests as diverse as those measuring formboard reconstruction and paper folding—but after extensive review of dozens of studies in which this factor has been identified, French (1963) recommended that a surface development test be added to the formboard and paper folding tests to measure this factor. This suggests that one could have defined this primary ability, before the advent of factor analysis, by choosing any one of the three reference tests and saying "this is it."

As for the representativeness of tests in factors, Tukey (1977) noted that

> when the right things can only be measured poorly, it tends to cause the wrong thing to be measured, only because it can be measured well. And it is often much worse to have a good measurement of the wrong thing—especially when, as it is so often the case, the wrong thing will in fact be used as an indicator of the right thing—than to have poor measurements of the right thing (p. 486).

The Educational Testing Service *Kits*, in short, while they provided researchers with reference tests for different cognitive factors, also contributed to a slowing down of new test development and to the growing tendency on the part of some psychologists to reify factors and treat them as if they represented significant domains of intelligence. By the mid-1970s, it was almost as if "to the statistician's dictum that whatever exists can be measured the factorist had added the assumption that whatever can be measured must exist" (Tuddenham, 1962, p. 516).

Lohman's Model

Lohman (1979) helped challenge this assumption by systematically re-analyzing factorial data from 35 previous studies using a common theoretical and methodological approach. He deliberately included in his analysis data that had either been obtained from studies using a variety of spatial tests (e.g., Thurstone, 1941; 1944; 1949); designed to clarify the nature of spatial ability (e.g., Guilford & Lacey, 1947; Michael et al., 1950); or which theorists had used to claim support for their models of factorial structure (e.g., Horn & Cattell, 1966; Guilford & Hoepfner, 1971). In several of his re-analyses, Lohman extracted oblique factors at several levels and then transformed factor matrices into orthogonal hierarchical structures using the Wherry procedure (Wherry, 1959). In some cases, both hierarchical and multidimensional scaling procedures were used (see Marshalek, Lohman, & Snow, 1983). Despite differences in the purposes of different studies, and despite an enormous diversity of tests and subject populations, Lohman found that a remarkably similar set of spatial factors appeared again and again.

In particular, he found three "major" spatial factors and a number of "minor" ones. Among the former, he defined spatial relations as a factor requiring the speeded rotation or reflection of figures or objects. Spatial relations was defined by tests like Thurstone's Cards, Flags, or Figures. Lohman noted that while mental rotation was the operation in common across these tests, spatial relations does not represent speed of mental rotation per se, but rather the ability to solve such problems quickly, by whatever means.

Lohman defined the factor of spatial orientation as representing the ability to imagine how a stimulus array would appear from different perspectives. This was a difficult factor to identify, however, because the same task could be solved by mentally rotating the array instead of re-orienting oneself. Tests that measure this factor are similar in intent to Piaget's and Inhelder's (1956) perspective tasks, but are usually presented as paper-and-pencil line-drawing tasks.

Visualization, the third major factor, was defined by a wide variety of tests such as Thurstone's (1938) paper folding test, French et al.'s (1963) hidden figures test, or Guilford's (1968) block rotations test. As Lohman noted,

> in addition to their spatial-figural content, the tests that load on this factor share two important features: (a) all are administered under relatively unspeeded conditions, and (b) most are more complex than corresponding tests that load on the more peripheral factors (Lohman, 1979, p. 127).

Lohman's "minor" spatial factors surfaced in his re-analyses only when highly similar tests were included in test batteries. These factors included *closure speed*, or the ability to identify quickly an incomplete or distorted picture; *perceptual speed*, or the ability to match visual stimuli rapidly; *visual memory*, or the ability to recognize a previously presented picture or geometric form; *spatial scanning*, or the ability to find a correct path visually in line-mazes or path-finding tests; *serial integration*, or the ability to integrate temporally visual stimuli from different locations; *flexibility of closure*, or the ability to break one gestalt and form another; and a *kinesthetic factor*, the ability to make left-right discriminations of various kinds. In a later paper, he observed that these "minor" factors differed to the extent that the tests that measured them varied in their complexity or speededness, in the type of mental operation(s) required for solution, and in the susceptibility of a given task to alternative solution strategies (Lohman, 1984, p. 19).

Earlier, Lohman concluded that, "at the most basic level, spatial thinking requires the ability to encode, remember, transform, and match spatial stimuli." He added that

> mental rotation, while an interesting and special type of mental transformation, is not the most important determinant of spatial ability. Rather, the crucial components of spatial thinking may be the ability to generate a mental image, perform various transformations on it, and remember the changes in the image as the transformations are performed. This ability to update the image may imply resistance to interference, both externally and internally. Further, it implies that one of the crucial features of individual differences in spatial ability may lie not in the vividness of the image, but in the control the imager can exercise over the image (Lohman, 1979, p. 116).

He also cautioned that "the central characteristic of spatial ability may lie in the nature of the internal representation rather than in the speed or efficiency of the various transformations applied to the image." Despite this caution, he proposed a hierarchical model that incorporated these dimensions of influence specifically.

Lohman (1984) argued that while the number of spatial factors may prove to be quite large, the number of processes required to solve existing spatial tests may be comparatively small. In particular, he claimed that only six processes are required, either singly or in some combination. These include pattern matching, image construction, storage (from iconic to long-term memory), retrieval, comparison, and transformation. Moreover, he claimed that there are two types of mental transformation. *Synthesis transformation* is most evident in tasks that require subjects to bring together or combine separate images; *movement transformation* is most evident in tasks requiring them to rotate, reflect, or transpose stimuli. As he noted,

> many complex tests require more than one transformation. Thus formboard (Ekstrom et al., 1963) requires synthesis, transposition, and rotation (p. 20).

Lohman's hierarchical model was an attempt to relate spatial tests and factors to different processes with respect to such influences as speed and task complexity. The vertical axis of his model represents the speed-power-complexity continuum, while the horizontal axis represents the nature and perhaps the complexity of the cognitive processes involved. Since test speededness is often confounded by test complexity, complex spatial tests in this model tend to cluster together near the top of the hierarchy, while simpler tests are fractionated into several specific factors at the base (see Figure 3.4 for a graphic representation). Further, factors nearer the base of the model are more sensitive to changes in test content or format, while tests for factors nearer the top are less sensitive to method variance.

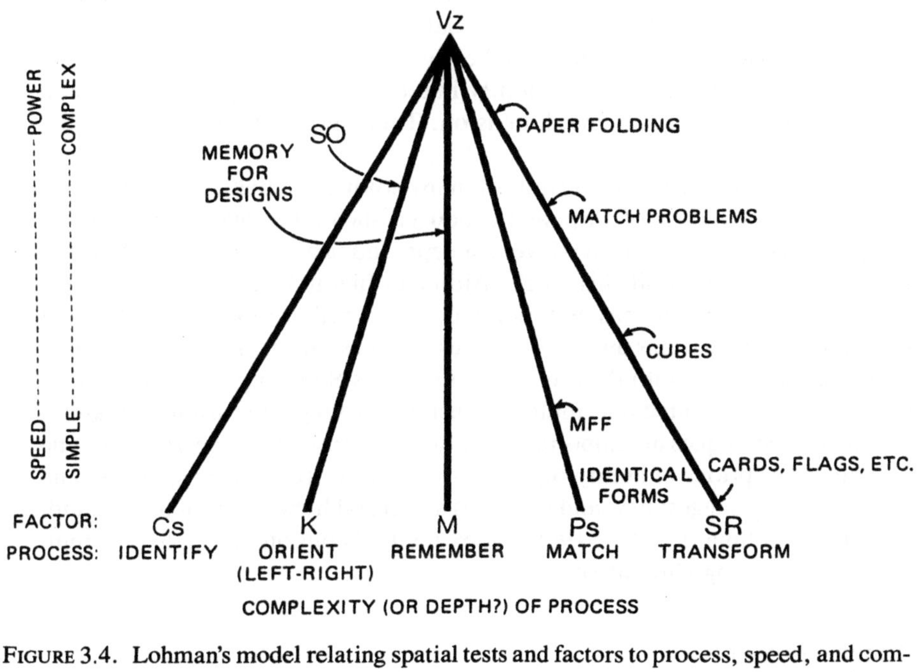

FIGURE 3.4. Lohman's model relating spatial tests and factors to process, speed, and complexity. (Reproduced with permission from Lohnman, D.F. [1979]. *Spatial ability: A review and re-analysis of the correlation literature*. Stanford, Stanford University Research Report, p. 132.)

As Lohman (1979) indicates, this representation lends itself to an explanation of a variety of factorial phenomena:

First, factors emerge only when individual differences in the particular processes required by tests can be elicited with sufficient strength to be reflected in the dependent measures that are employed. For example, individual differences in the number of pictures correctly identified is elicited by degrading, distorting, or erasing part of the picture, as in the various Closure Speed tests.

Second, task complexity may be increased by increasing either the number of distinct operations, or the difficulty of each operation. Thus, tasks which elicit individual differences in memory and transformation should be more complex, and thus produce a factor further up in the hierarchy than comparable tasks where individual differences in one component are elicited. On the other hand, task complexity may be increased by increasing the difficulty of the component that produces individual differences (p. 131).

Lohman's model is only one of several possible examples of hierarchical models. It has been described here because it is domain-specific and does not attempt to account for spatial abilities within a structure or organization of all known intellectual abilities. It also differs from other models such as Guilford's in that it is generated from performance on existing paper-and-pencil spatial tests, and does not suggest the development of new spatial tests per se. On the other hand, it is derived from one of the first large-scale re-analyses of data from previous studies, and is also one of the first to include such dimensions of variance as speededness and item complexity. As such, it may be a rich source of future hypotheses about the nature of spatial thinking. An example of such an application is provided by Pellegrino and co-workers (1984), who extrapolated a two-factor representation of spatial ability in terms of a speed-power dimension and a simple-complex processing dimension. Their representation is presented graphically in Figure 3.5.

In the most recent decade, the results of many studies of intelligence have been factored in such a variety of ways and produced such a bewildering number of primary-, secondary-, and tertiary-level concepts and factors, that it is difficult to communicate a sense of all this work. Although hierarchical models of factor structure have become increasingly popular, other, rather different methods of analyzing the interrelations among measures, such as multidimensional scaling, have produced models with different structures. As Snow, Kyllonen, and Marshalek (1983) have pointed out, multidimensional scaling and various hierarchical clustering techniques are important because their results may corroborate one or another conceptions of ability organization derived from factor analysis, and because they may suggest new and different structural hypotheses of use in guiding research. Guttman's (1954; 1965) radex model is but one example of a multidimensional scaling alternative.

Guttman's Model

Guttman's model originated in his work with factor analysis, but gained its main support from his non-metric scaling analysis of portions of Thurstone's (1938;

1941) work. The model itself was derived using non-metric scaling techniques and a faceted definition of the universe of ability tests. As Guttman (1954) described it:

Within all tests of the same kind, say of numerical ability, differences will be in degrees. We shall see that addition, subtraction, multiplication, and division differ among themselves largely in the degree of their complexity. Such a set of variables will be called a simplex. It possesses a simple order of complexity. The tests can be arranged in a simple rank order from least complex to most complex.

Correspondingly, all tests of the same degree of complexity will differ among themselves only in the kind of ability they define. We shall postulate a law of order here too, but one which is not from "least" to "most" in any sense. It is an order which had no beginning and no end, namely, a circular order. A set of variables obeying such a law will be called a circumplex, to designate a circular order of complexity. Our empirical data will testify that different abilities, such as verbal, numerical, reasoning, etc. do tend to have such an order among themselves (p. 260).

Put differently, Guttman's radex model brings together the idea of a difference in kind between tests and the idea of a difference in degree:

tests are represented as points in two (or three) dimensional space; the higher the correlation between the tests, the greater their relative proximity in the space. Guttman observed that the ability tests *within* a content area (spatial, verbal, or numerical) tend to form a simplex, a straight line array in scaling representation, on which tests are ordered from simple to complex. Tests of comparable complexity sampled from separate content areas tend to form a circumplex, a circular array in scaling representation. The covariation of the simplex and circumplex structures was hypothesized to form a radex—a disc in two-dimensional space or a sphere in three-dimensions, divided into verbal, numerical, and figural content areas (Marshalek, Lohman, & Snow, 1983, p. 107).

Figure 3.6 is a schematic representation of one form of radex sphere, which emphasizes a continuum on which spatial tests differ in the number of performance components required. Guttman predicted that complex tests would fall on

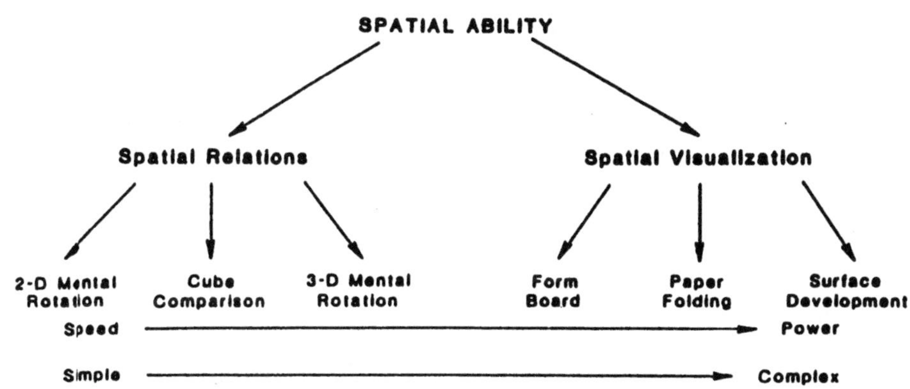

FIGURE 3.5. Schematic representation of spatial ability. (Reproduced with permission from Pellegrino, J.W., Alderton, D.L., & Shute, V.J. [1984]. Understanding spatial ability. *Educational Psychologist, 19*, 243.)

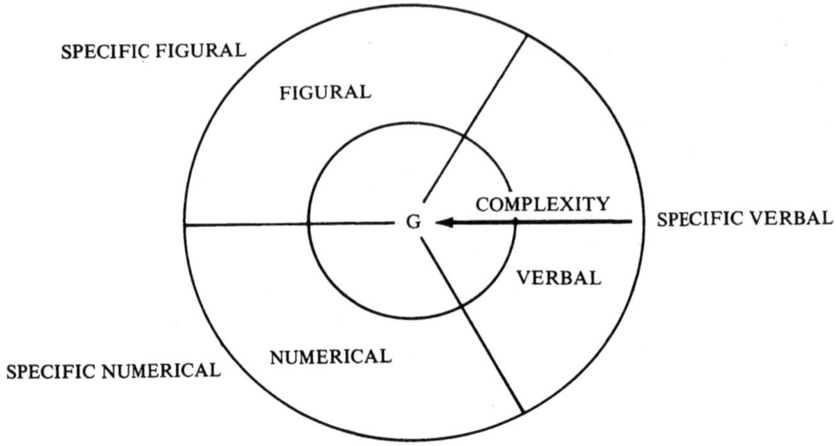

FIGURE 3.6. Schematic representation of Guttman's radex sphere. (Reproduced with permission from Marshalek, B., Lohman, D.F., & Snow, R.E. [1983]. The complex continuum in the radex and hierarchical models of intelligence. *Intelligence*, *7*, 110.)

the periphery of the radex sphere because complexity could result from many different combinations of components. He assumed, in other words, that complex tests would have fewer components in common with each other than would simpler tests, and thus would show lower intercorrelations (Snow, Kyllonen, & Marshalek, 1984).

Although radex structures may be obtained by other methods, multidimensional scaling best accommodates the study of content facets. Guttman and Shoham (1982), for example, administered eight spatial tests to 800 subjects in order to test whether the structure of intercorrelations between tests conformed to the structure hypothesized on the basis of a mapping sentence. Data were first analyzed by calculating an intercorrelation matrix between items from each of the tests, and then by a Smallest Space analysis to ascertain whether the content facets corresponded to regional partitions in space.

Guttman and Shoham selected eight spatial tests to represent five of eight possible structuples from three dichotomous facets of content. Their content facets were 1) the *type of rule* required for solution; 2) the *stimulus dimensionality* of test content; and 3) whether the imagery for solution was *static or kinetic*. A further facet was the specific test given. Their mapping sentence read: "The success of person (x) in a cognitive task of 1) inferring or applying a spatial rule to 2) stimuli of two- or three-dimensional content using 3) static or kinetic imagery to 4) respond to a particular kind of spatial problem results in high/low success in solving cognitive spatial tasks."

In order to construct the mapping sentence, Guttman and Shoham surveyed the literature on spatial abilities to arrive at content facets. Specifically, the *type of rule* was taken from the radex model of intelligence advanced by Guttman (1965); *stimulus dimensionality* of test content from Vandenberg (1969), Yen (1975), and McGee (1979); *static or kinetic imagery* was from Guilford and Lacey (1947), and Thurstone's (1951) Sl factor; and *format* referred to the particular test (e.g., Hidden Patterns, Progressive Matrices etc.). Type of rule required for solution was defined in terms of whether the subject must find the rule needed to solve the test problem (rule inference) or whether she must simply apply the rule of a given relationship. Items on the Raven's Progressive Matrices, for example, require rule inference since the subject must determine the logic of a pattern and then make an inference as to which piece will complete the pattern according to the rule. The Hidden Blocks test, by contrast, is a rule application test because the subject is told to find the number of blocks in a pile and must "apply" this rule by counting blocks.

Figure 3.7 is a two-dimensional projection of a three-dimensional space, showing a partitioning into regions corresponding to all facets in the mapping sentence. Tests requiring rules of inference are to the left of an arbitrary line in the figure; application tests are on the right. Tests requiring mental rotation of figures are all located within the outline of the broken line. Raven's Progressive Matrices, The Spatial subtest of the Differential Aptitude test, the Thurstone Hidden Block and Hidden Patterns tests are on the outside. Except for Llana Rotations (similar to Shepard and Metzler's 1971 block rotation task) and the

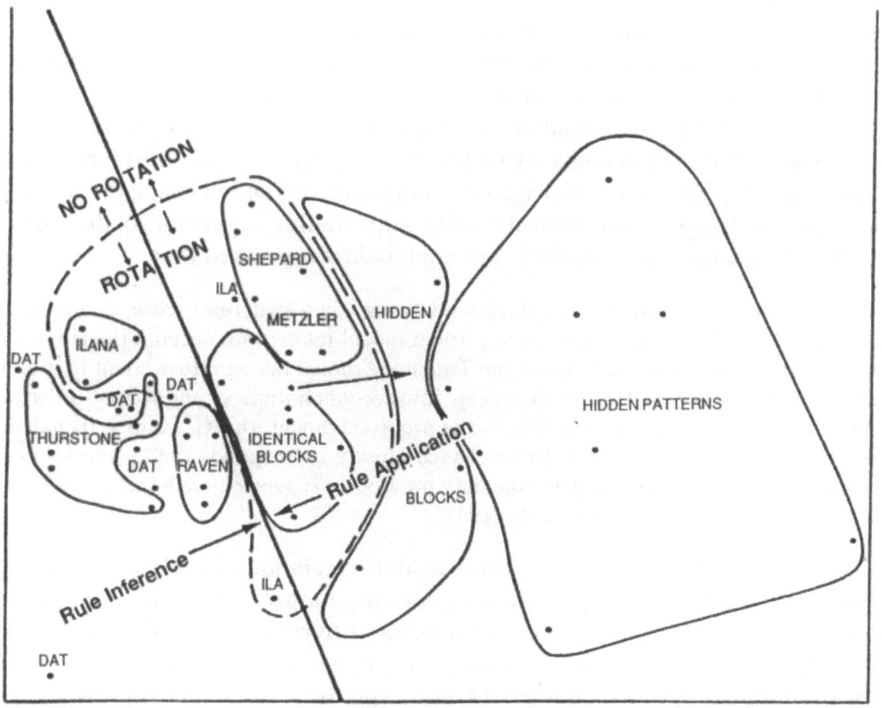

FIGURE 3.7. Guttman and Shoham's partitioning into regions that correspond to facets from a mapping sentence. (Reproduced with permission of the authors and the publisher from Guttman, R., & Shoham, I. The structure of spatial ability items: A faceted analysis. *Perceptual and Motor Skills*, 1982, *54*, 487–493, Figure 1.)

spatial subtest from the Differential Aptitude Tests, the format facet provides a clear partitioning in the sense that items of the same test tend to fall into the same region of space. Guttman and Shoham see the task of future researchers as refining and extending such knowledge about the structure of spatial abilities.

Even more recently, Gustafsson (1981; 1982; 1984) reported a study in which 1224 sixth-grade children were administered a battery of 13 ability tests, seven of which were known to load upon one or another spatial factor. Using the LISREL technique developed by Joreskog (1978; 1981), he found support for primary factors in the Thurstone tradition, as well as for second-order factors similar to those described by Horn and Cattell (1966). Gustafsson also decided that a second-order factor of fluid intelligence was identical to a third-order (G) factor. On the basis of these findings, he proposed a three-level hierarchical model which, he claimed, incorporated the best features of previous hierarchical models.

Gustafson's HiLi-Model

Gustafsson described his "HiLi-model" as a general model of the structure of human abilities, of which most other models could be viewed as special cases. His name for it is an acronym for *Hi-erarchical Li-srel* model. The characteristics of this model, as the name indicates, are that it is hierarchical, it is based upon LISREL techniques, and it includes a set of basic factors such as (G), and Cattell's (Gc), and (Gv). The number of levels in the hierarchy may vary, as may the domains of tasks that are sampled. When a certain domain is represented by only a few tests, there will necessarily be few levels in that portion of the hierarchy.

Gustafsson (1984) cautioned against comparing and emphasizing the similarities between different hierarchical models until studies have been conducted that include a broad range of abilities. He nonetheless suggested that

> the Spearman, Thurstone, Cattell-Horn models may, in a structural sense, be viewed as subsets of the HiLi model: with the Spearman model taking into account the amount of variance from the third-order factor; the Thurstone model taking into account first-order variance; and the Cattell-Horn model taking into account both first- and second-order variance. The Vernon model comes close to the proposed model: the 'G' factor is included in both models, and the second-order level v:ed closely corresponds to Gc, and k:m corresponds to Gv. Guilford's model is the only previously suggested model which is clearly incompatible with the HiLi model (p. 193).

According to Gustafsson, the factor analytic techniques used in these earlier studies were flawed because of their *exploratory* nature, and because they failed to estimate the relative influence of factors at different levels. LISREL, by contrast, is a method of analysis for estimating and testing *confirmatory* models. In confirmatory models, the number of factors and the patterns of their loadings are known in advance from previous research. The HiLi-model, in other words, represents an advance in our ability to estimate and test hierarchical models of intelligence.

Gustafsson began with 20 first-order factors and ended with a (G) factor. The HiLi-model is graphically represented in Figure 3.8. Tests are depicted in enclosed squares; factors in enclosed circles. Straight arrows indicate direction of influence, and bi-directional arrows refer to correlations without assumed causality.

The complete dependence of the HiLi-model upon tests with statistically demonstrated factorial loadings restricts its usefulness to clarifying what is known about the structure of abilities when measured by existing tests. Such a restriction is a useful reminder that in factor analysis one gets out of the analysis only what one puts into it—and that test development over the past forty years has become remarkably reiterative. The fact that several hundred spatial tests can be grouped into ten single-task categories (see Chapter 1), is testimony to this circularity. It also raises serious questions about the adequacy with which this comparatively limited range of test categories represents the domain of intelligence known as spatial ability.

Although there are a number of other recent psychometric models that include spatial ability as a factor in their organization of abilities, perhaps Carroll's hierarchical model best represents these efforts.

Carroll's Model

Carroll's (1985) model is based upon the systematic re-analysis of approximately 300 factor data sets and, although his work is not yet complete, he has reported about thirty different factors or sources of variance that appear to constitute a hierarchical ordering. One of these is a second-order factor of general visual perception, including many of the spatial factors which Lohman (1979) described in his analysis, and a third-order factor which Carroll labeled general intelligence or (G). A graphic representation of Carroll's model is presented in Figure 3.9.

This model is an ambitious undertaking which Snow (1986) believes to have significant implications for educators. Carroll hypothesizes that the higher-order factors do not necessarily subsume lower-order factors, but simply differ in their range of applicability to cognitive tasks. Thus, as Snow (1982) observes,

> visual-spatial abilities appear to pop in and out of relevance as spatial reasoning and other operations with visual images are required by chosen materials, problems, or forms of teaching. Visual-spatial abilities have been associated with achievement in art, architecture, dentistry, and technical courses such as shop, mechanics, or engineering design (p. 542).

To summarize this most recent phase of spatial measurement, psychologists using the psychometric approach can be said to have substituted the re-analysis of previous data for the undertaking of large-scale data collection; have shifted their interest from multiple-factor models of intelligence to a variety of hierarchical ones; have increasingly depended upon a limited range of existing tests with known psychometric characteristics rather than developing new ones, and have

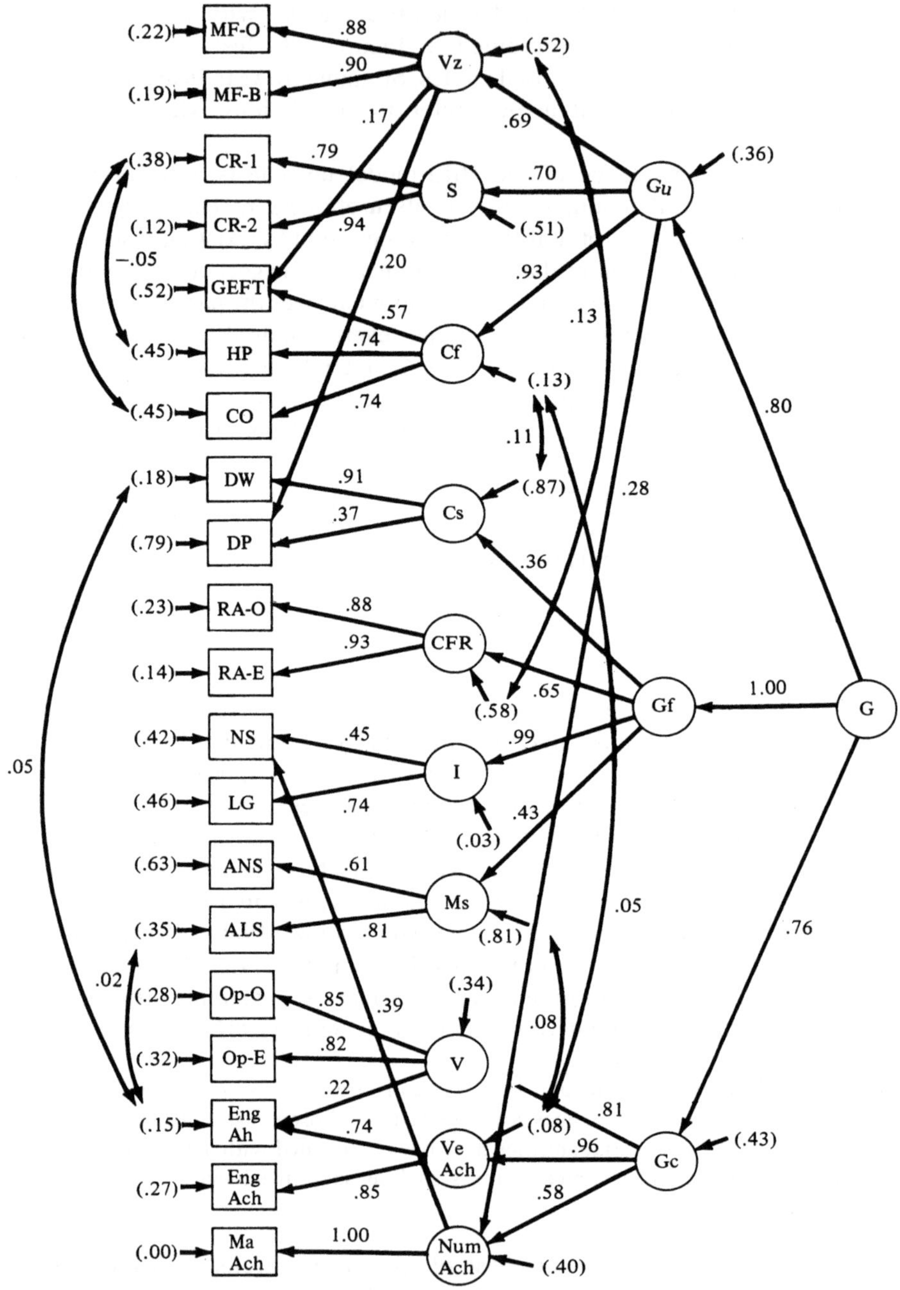

FIGURE 3.8. Gustafsson's hili model. (Reproduced with permission from Gustafsson, J.E. [1984]. A unifying model for the structure of intellectual abilities. *Intelligence*, *8*, 192.)

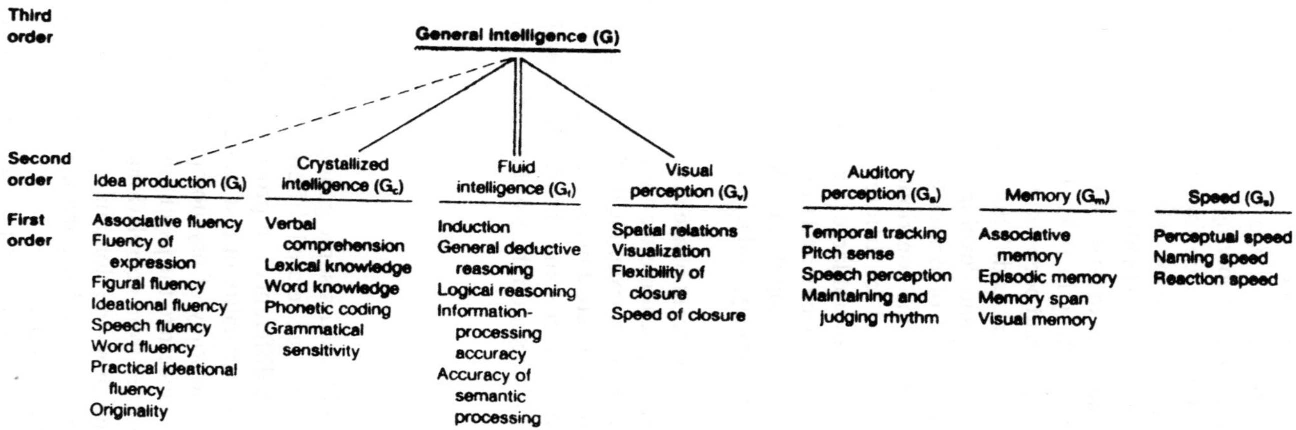

FIGURE 3.9. Carroll's hierarchical model. (Reproduced with permission from Carroll, J.B. [1985]. *Domains of cognitive ability.* Paper presented at AERA Conference, Los Angeles, California, May 1985.)

become more and more preoccupied with clarifying and confirming what is known about the organization of human abilities.

In all three phases described here there have been important advances in theory, methodology, and test development which have contributed to our understanding of spatial ability. Spearman's two-factor theory of intelligence, based upon a complicated methodology involving the use of Tetrad-difference equations, dominated psychometric research for more than thirty years. The very gradual identification of a spatial factor over and above a general ability factor can be ascribed to limitations inherent in his methodology; to the lack of adequate tests; and to an overwhelming bias in favor of verbal abilities. With the introduction of multiple-factor methodology in 1931, however, psychometricians found that the factor of general ability had been rotated away and replaced by a number of independent primary abilities instead. Thurstone deserves credit for introducing this alternative methodology and developing many new tests, as well as for using both to support his model of primary mental abilities. World War II provided the opportunity to use his contributions in large-scale research programs to screen military personnel for different assignments.

The second phase witnessed several large-scale efforts to establish differences between subdivisions of a spatial factor and the development of many new paper-and-pencil instruments for measuring them. This phase ended, however, with a confusion of factors and terminology, and a turning away from a focus upon spatial abilities per se.

As we have seen, in the most recent phase there has been a pronounced growth of both number and kind of psychometric methodologies, of the models proposed, and of studies undertaken to examine different sources of variance. Many such studies were prompted by interest in general and specific issues with respect to spatial ability. We shall consider some of these issues and associated research as a way of concluding this chapter on the psychometric approach to the study of psychological space.

Issues

As indicated earlier, we will discuss some general issues such as the emergence and decline of spatial ability, sex-related differences, the trainability of spatial ability, the role of imagery, differences in processing spatial problems, and the effects of the stimulus dimensionality of test content on performance, which cut across all approaches to the study of psychological space. We will also look at some issues of particular concern to the psychometric approach, such as the numbers of factors to be extracted from correlational matrices, when the rotation of factors should be stopped, how the domain of spatial abilities relates to other domains in various structures of ability, and the effects of speededness and task complexity upon spatial performance.

With respect to the emergence and decline of spatial ability, it must be emphasized that there have been very few studies where spatial ability by itself has been the focus of longitudinal comparisons. The stability of factors over different ages has also been difficult to measure because we lack tests that are known to require similar solution responses at different age levels. Despite these difficulties, Reinert (1970) found over 60 publications on comparative factor analytic studies of intelligence across the life-span – a number that has grown considerably since 1970 (see Dixon, Kramer & Baltes, 1985).

As with other general issues, it is not possible here to review all theories and empirical evidence. However, we can add some contextual information to the differences in viewpoints already described.

Although Binet's and Simon's *Scales of Intelligence* (1905) were deliberately developed as a progression of age-related tasks, there were comparatively few additional studies of the stability of factors over age during the first phase of mental testing. In 1946, however, Garrett summarized the early efforts and also proposed what Anastasi (1958) later called the *age-differentiation* hypothesis:

> Abstract or symbolic intelligence changes in its organization as age increases from a fairly unified and general ability to a loosely organized group of abilities or factors (Garrett, 1946, p. 373)

Vernon criticized this hypothesis because Garrett had based it upon comparisons between college and high school students and high school and elementary school students. He suggested that the smaller (G)-variance in the older groups might simply reflect their greater selectivity (Vernon, 1950, p. 29).

Burt (1949), in his review of the contribution of practical ability to success in trade or technical schools, concluded that practical ability becomes manifest at about age 11, increases in importance until about age 15, and becomes more specialized thereafter. In 1954, Burt reviewed data in support of Garrett's hypothesis.

If the age-differentiation hypothesis was advanced to explain changes in intellectual abilities during childhood, Horn's and Cattell's (1966) hierarchical theory enabled psychologists to consider changes across the entire lifespan. In particular, their alternative version of "the classic aging pattern" was based upon their distinction between fluid (gf) and crystallized (gc) intelligence:

> Crystallized intelligence, because it is indexed by the life-long accumulation of cultural knowledge, usually increases over the adult years. Fluid intelligence is more dependent upon physiological functioning and especially on the support of a relatively determinate neurological base. If this neurological base, which is continually subject to change, is impaired, the ability to perform the underlying intellectual skills is undermined. Furthermore, the growth of the cognitive competence associated with crystallized intelligence may also contribute to the eventual age-related decline in fluid performance levels. Thus the functional prediction is that, from a peak in late adolescence, fluid intelligence will decline throughout the adult years, whereas crystallized intelligence will evince progressive increases in performance throughout adulthood (Horn, 1982, p. 320).

As spatial reasoning tests are markers for fluid intelligence, it would seem reasonable to expect an inevitable age-related decline in spatial intelligence in adulthood (ibid., p. 864).

The latter expectation prompted a series of studies to investigate the psychometric variability of older adults (e.g. Willis, 1980; Baltes & Willis, 1982). As an example, Willis, Bliezner, and Baltes (1981) trained a group of adults of about 70 years of age to use verbal procedures to solve problems requiring the manipulation of figural relations. A variety of (gf) and (gc) ability measures were then administered in order to test transfer of training after one week, one month, and six months. Compared with adults in a control group, the trained subjects demonstrated both specific and generalized transfer effects over the six-month interval.

Without going into the controversy about methodologies suitable for assessing lifespan changes in intelligence, it is worth noting the longitudinal studies of Schaie (1956; 1968; 1983) and his comparison of inferences drawn from research that he reported in 1970. According to Schaie, cross-sectional studies show a sharp decrement in spatial abilities from a peak at young adulthood to old age; longitudinal studies show a modest decrement from an adult plateau, and short-term longitudinal studies show almost no decrement until advanced old age (Schaie, 1970, p. 499). Clearly more research is needed before it is possible to make strong statements about age-related changes in spatial ability or in factor structures across the life-span.

As indicated earlier, the issue of sex-related differences in performance on spatial tests has been a matter of controversy ever since Maccoby and Jacklyn (1974) published their review. McFarlane (1925) was one of the first to report such differences with her performance test, and Werdelin (1961) was one of the first to comment upon possible differences in processing to account for his finding of sex-related differences on a battery that included spatial tests. Whereas a large proportion of psychological research during the late 1960s and early 1970s was focused upon age-related differences (thanks in part to Piaget), during the late 1970s and early 1980s it could be said that this interest was supplanted by a fascination with sex-related differences as far as spatial abilities were concerned. Coincidental with the enactment of the Equal Opportunity Employment Act, Buffrey and Gray (1973) noted that only about one girl in twenty performed as well as the average boy on putative spatial tests requiring the mental manipulation of spatial relationships. As Bejar (1985) noted, it was ironic that with the increase of concern about civil rights and job-related discrimination came a decrease in the amount of large-scale testing for classification purposes, especially when involving spatial tests.

Although the fact of sex-related differences is well documented, the extent of such differences and the explanations for them have been the focus of considerable debate. Although some psychologists have said that the presence of sex-related differences may be a criterion for characterizing a "spatial" test (Eliot and Smith, 1983), others have insisted that the size and consistency of these differences have been overestimated (Caplan, MacPherson, & Tobin, 1985). With respect to the latter view, Kimball observed that

> the observed differences are very small, the overlap large, and the abundant biological theories are supported with very slender or no evidence (Kimball, 1981, p. 333).

With respect to explanations for these differences, some researchers have advanced "biological" theories, including the idea that differences are transmitted by an X-linked recessive gene (e.g. O'Connor, 1943; Stafford, 1961; Fralley, Eliot, and Dayton (1978) and Thomas, 1983), that they reflect the effects of chromosomal and metabolic influence upon cognitive functions (Broverman et al., 1968), or that they are related to maturational differences in the brain (Waber, 1979; Harris, 1978). As Caplan et al. (1985) have pointed out, some difficulties associated with these theories include

> ambiguous and contradictory findings, faulty extrapolations from data that seem only peripherally related to spatial abilities, bias in selecting variables, faulty logic, and inconsistent reports (p. 790).

Other researchers have argued that sex-related differences reflect critical differences in childrearing, learning opportunities, or cultural expectations (see Sherman, 1978). And still others maintain that they are a consequence of a complex interaction of biological and social factors (e.g., Bock & Kolakowski, 1973).

Linn and Petersen (1985) undertook a meta-analysis of group mean differences from a variety of tests, and identified sex-related differences for three aspects of spatial ability. They report that spatial perception and mental rotation tasks were easier for males than for females, but that spatial visualization

> which is characterized by analytic combination of both visual and non-visual strategies, is about equally difficult for males and females (p. 1491).

As they acknowledge, the results of their meta-analysis are only as valid as the empirical data from the studies entered into it.

Another example is Hilton's (1985) comparison of scores from tests that were administered to high school students in *Project Talent* in 1960, and in *High School and Beyond* in 1980. Hilton reported that while mean scores for all tests in the battery declined over the twenty-year interval, the mean score for the spatial test declined the most. Moreover, sex-related differences on the spatial test became noticeably less in the 1980 sample. Hilton attributed this differential decline to a convergence in the experiences of male and female subjects.

These last two examples are mentioned simply to indicate that the controversy over sex-related differences is still very much alive today. As Caplan et al. (1985) observe, the controversy persists in large part because there is no widespread agreement about the characterization of spatial ability. From the psychometric viewpoint, it is difficult to clarify its nature when test development over the past forty years has been, for the most part, remarkably reiteratively descriptive.

Evidence for age-related and sex-related differences inevitably raises questions about the trainability of spatial intelligence. There are many instances in the literature wherein results from spatial training programs claim that educationally significant improvements have been effected. Unfortunately, most of these studies have not examined the matters of retention and transfer. Moreover, even when these criteria are applied, many studies are difficult to interpret because of inadequacies of design, problems with methodology, or limitations in measures used to assess retention and transfer.

The issue of trainability has been studied using three types of investigations. Early studies entailed a pretest posttest design, with training consisting mainly of some experience thought to require a high degree of spatial ability. Later studies gradually became more controlled as researchers sought to manipulate particular variables, and recent research has looked at differences in the training of high- and low-ability subjects upon specific spatial tasks.

Among the early studies, Churchill, Curtis, Coombes, and Hassell (1942), on the one hand, found a slight but significant gain in performance on a surface development task by a group of engineering students who had spent nine weeks in a draftsmanship course. On the other hand, Faubian, Cleveland and Hassell (1942) found no such gains on a similar test when Air Force recruits were given training in draftsmanship and blueprint reading.

In a later study, Krumboltz and Chrystal (1960) assigned 512 Air Force ROTC students to four groups. Subjects in the first group were given the same spatial pre- and posttest. The second group was given an alternative form of the spatial pretest as their posttest; the third group was given different pre- and posttests, and the fourth group received only the pretest. Different intervals (10 minutes or 7 hours) between pre- and posttesting were not found to have an effect, but subjects in the first and second groups showed significantly greater training effects.

A more recent example is Rowe's (1982) training study. Rowe attempted to assess the effects of two training programs upon 200 seventh-grade subjects. She gave one group a program consisting of tasks that required two-dimensional thinking and the forming but not manipulation of visual shapes. A second group received a program containing tasks that required three-dimensional thinking and often the mental manipulation of images. Both treatment groups outperformed a control group on retest performance after 20 one-hour training sessions over a 23-week period, and again after a year's interval.

Among her findings, Rowe reported that the two-dimensional training program significantly affected subjects' performances on spatial tasks across all ability levels. However, subjects in the three-dimensional training program, while they improved on two-dimensional tasks, were less successful with three-dimensional ones when subjects were of low ability.

Results from most training studies, while often encouraging, are also often inconclusive. When subjects are trained indirectly through courses that are thought to require high levels of spatial ability, the results have been mixed. If they are given practice on tasks that are highly similar to the tasks on spatial tests, the results can be interpreted as indicating that spatial ability, as a set of operations, can be trained. At the present time, however, we do not have results from long-term training efforts to indicate whether training effects persist.

A fourth general issue, the role of imagery in spatial thinking, has had a long history in psychology. Galton, for example, thought that "a visual image was the most perfect form of representation wherever the shape, position, and relations of objects in space was concerned" (Galton, 1883, p. 113). When El-Koussy (1935) described his (K) factor, he said that he thought it represented "an ability to obtain and a facility to use visual spatial imagery." And when Thurstone (1938) described the distinguishing characteristic of tests loading upon his space factor, he thought it was "a facility in holding a mental image and mentally twisting, turning, or rotating it to a different position and matching it with a suggested solution." Although the nature of visual imagery was not defined by these early psychometricians, they clearly assumed that it was intimately associated with spatial ability.

Despite their alleged intimacy, the relationship between imagery and spatial ability is still not well understood. Although factor analysts have repeatedly described spatial ability as entailing the use of visual imagery, there have been few efforts to develop models that have related specific spatial factors to different imagery processes or structures. A difficulty in developing such models is the fact that the terms "visual imagery" and "visualization" are often used interchangeably, and the fact that tests that have been devised to measure both constructs are often one and the same test.

One of the few attempts to explore the relationship between spatial ability and imagery was undertaken recently by Poltrock and Brown (1984). These investigators administered eight spatial tests, three imagery questionnaires, and six laboratory tasks to 79 subjects in an effort to ascertain the relationship between performance on spatial tasks and their subjects' image quality and image process efficiency. The laboratory tasks were designed to assess image generation, image scanning, adding and subtracting detail in images, and image integration. The eight spatial tests loaded upon a single spatial factor, visualization.

Although ratings of imagery control and vividness of imagery were found to be unrelated to spatial performance, the measures of process efficiency and image quality were strongly related to spatial performance but weakly related to each other. Poltrock and Brown advanced a structural equation model which indicates that spatial visualization ability may be decomposed into a unique variance plus a linear combination of different imagery measures. They suggest that their model indicates the most important determinant of spatial visualization is not rotation efficiency, but the maintaining of a high quality image of the stimulus (Poltrock & Brown, 1984, p. 136).

Lohman (1984), when summarizing dimensions of individual differences in spatial abilities, pointed out that most spatial tests require abstract figural problem-solving skills. He argues that "we probably come closer to understanding performance on most spatial tests by understanding reasoning and problem solving than by understanding visual imagery." Moreover, he suggests that spatial ability may not consist so much of the ability to transform imagery as to create the type of abstract, relation-preserving structure on which transformations may be performed. It is interesting to note how closely this suggestion resembles Galton's (1883) idea about the role of visual imagery in spatial thinking!

A fifth general issue, the types of processing required by different spatial tasks, has been addressed in a variety of ways. Kelley (1928) and Thurstone (1938) distinguished between tasks that required the sensing and retention of forms (perceptual processes), and those that required the manipulation of spatial relationships (conceptual processes). A somewhat different expression of this issue is the distinction between spatial tasks that appear to require analytical or synthetical processing for their solution. According to Smith (1964), French (1965), Guay, and McDaniel (1978), Lohman (1979), Willis (1980), and others, analytical processing entails the successive or sequential trial-and-error checking among the various parts of a stimulus, while synthetical processing entails the simultaneous or holistic response to the stimulus as an organized whole. Some psychologists, like Smith (1964), insist that only the tasks requiring synthetic processing are truly spatial and, by implication, if tasks do not elicit this strategy, they cannot be considered "true" tests of spatial ability.

It has been suggested, of course, that the same task may elicit a number of processing strategies. P. E. Barratt (1953), for example, claimed to have identified four different strategies used to solve a space relations item. Allen (1974) reported from 14 to 20 possible strategies (from task specific to general) for tests requiring cube comparisons and paper folding. If the process used to solve a task also determine what ability the task measures, many so-called spatial tests may actually be measuring distinctly different abilities for different people. Certainly, the susceptibility of some spatial tests to non-spatial strategies serves to confound measurement of spatial ability.

Lohman (1979) pointed out that individual differences in processing or solution strategies challenges a basic assumption of factor analysis. As he observed,

> Factor structures obtained from analyses of test scores may be severely distorted. The most likely outcome is an overestimation of the factorial complexity of a test. Thus, that some spatial orientation tests load on both a visualization and a spatial orientation factors may only mean that students solve the tests differently; some use a predominate spatial orientation strategy while others rely upon a visualization strategy. Alternatively, students may switch between these two strategies while solving different items. However, even in this straightforward example, it is impossible to know whether the test measures different aptitudes in different individuals. On a more general level, the presence of several tests in a battery that are amenable to alternative solutions strategies seriously distorts the factor structure, so that the obtained factor may not apply to anyone in the sample (p. 191).

He concludes that the challenge for future research is to devise experiments that reveal solution strategies for each subject on every item or item-type.

The last general issue is the effect of stimulus dimensionality of test content on factor loadings. This issue surfaced, as we have seen, with the development of performance and group tests during and after World War I. Thurstone (1941), Moffie (1940), Burt (1949), Guilford (1967), Ho (1974), and Guttman and Shoham (1982) are among those who either have questioned whether such an effect exists or have attempted to ascertain its nature empirically.

Moffie, for example, paired two-dimensional Thurstone tests with similar three-dimensional performance tests (e.g., paper formboard with wooden formboard). Although no dimensional differences were found, this study drew attention to the possibility that stimulus dimensions of a given test might affect an individual's performance.

The issue of stimulus dimensionality also surfaces as questions about the adequacy of existing tests of spatial abilities. Currently, it is unclear whether tests of figural, object, or large-scale space are measuring the same underlying construct. If measures for different spaces do not have a common factor structure, then we will be forced to talk about a plurality of psychological spaces, and ascertain what tests will be required to assess situations where all three spaces (figural, object, and large-scale) are perceived and represented simultaneously.

With respect to some issues that are specific to the psychometric approach, several of these reflect limitations inherent in the use of correlational or factor analytic methodology. As we have seen throughout this summary, controversy persists both with reference to how many factors should be extracted from correlational matrices, and to decisions about stopping the rotation of factors. These controversies are manifest in such issues as how many spatial factors or subfactors exist, and how they relate to one another. They arise in such questions as how spatial abilities are related to other perceptual, visual, or non-verbal abilities in various structures or organizations of intellectual activities. Although some psychometricians treat spatial ability as an information-processing dimension, others like Guttman (1980), Gustafsson (1984), and Carroll (1985) treat spatial ability as a domain of intelligence in its own right. It may be that with the development and application of newer methods of analysis, or with the further re-analysis of existing factorial data sets, psychometricians will find sharper boundaries between factors and clearer understandings of their complex interrelationship.

Another specific issue is the extent to which test constraints such as speededness and item complexity affect the loadings upon various spatial factors. As will be discussed further in connection with the experimental approach, the relationship of test constraints upon the subject's choice of processing or solution strategy is still very much a matter of continuing controversy.

A final specific issue is the extent to which abilities are simply reflections of different task demands. To paraphrase Neisser (1976), tests of spatial ability typically involve the recognition or manipulation of symbolic information in order to solve tasks for which all necessary information is provided. Most tests are designed to evince one correct answer for each item, a limited amount of time for subjects' to reach a solution, and items that are remarkable for their minimal relevance to real life. As Stea and Blaut (1973) observe, subjects are frequently given tests that require particular abilities or capacities so that researchers may demonstrate, in turn, the characteristics of those same abilities or capacities. Since researchers look at test results in order to discover what the tests are measuring, we cannot always be certain that the abilities or capacities which a theorist ascribes to a factorial model are not artifacts of the methods we are using to generate or elicit those same behaviors (also see Gould, 1981).

In this chapter, the psychometric approach to the study of psychological space has been described in terms of three phases of the mental testing movement. Within each phase, a number of different models and studies were described, first with respect to efforts to identify a group factor of spatial ability over and above a general ability factor, next with respect to the proliferation of spatial factors and efforts to understand their relation to one another, and finally with respect to efforts to identify different sources of variance in performance on spatial tests, and to relate spatial abilities to other intellectual abilities in factor structures. Following this summary, six general issues and a number of specific issues were discussed briefly. In the next chapter, a similar range of models and issues will be described from the developmental approach.

4
The Developmental Approach

The psychological study of human development is primarily the concern of those who assume age-related differences to be important both in terms of when they occur and how they present themselves behaviorally. Typically, the developmental psychologist studies individual differences by presenting subjects with specific tasks or problems, and by examining their responses in terms of hypothesized stages or age-related changes. Individual differences, however, are not the primary concern of developmentalists: their major efforts are usually directed to the study of *inter-individual* differences, with chronological age as the important variable, or some other variable that correlates with age (e.g., grades in school). When individual differences are studied, they are usually interpreted in terms of the achievement or acquisition of *universal patterns of behavior*, and models—especially models of intellectual development—are most often advanced to account for changes in terms of *stage characteristics of knowing*.

The study of spatial development is only one of several domains of behavior within human development generally and intellectual development in particular. Developmentalists interested in psychological space usually ask questions about whether systematic shifts in spatial functioning occur with age, whether these shifts are meaningful in relation to other processes, what variables affect these shifts, and whether differences in spatial functioning occur both within and across age levels. The developmentalist identifies tasks or problems as "interesting" if they possess physiological or neurological relationships, if they possess qualitative as well as quantitative dimensions, and if they can be linked in occurrence with a set of causes. While a variety of statistical methods are employed, there is a tendency to depend upon analysis of variance in studies that emphasize chronological age as the important variable.

Within the ongoing study of human development, the developmental study of psychological space has had a very uneven history. Problems of spatial perception and representation were of tangential interest to most psychologists for the first sixty years of this century. Then, with the shift of interest in the 1960s from behavior to complex internal processes, there came a dramatic increase in the amount of attention paid to questions of spatial development, in terms of both theoretical advances and research activity. More recently, there has been an overall decline of research activity concerned with spatial development in childhood, and a broadening of interest in studies of spatial behaviors across the lifespan.

Although it is tempting to begin our discussion of spatial development with Piaget's model and a reference to research activity of the 1960s, this would present a distorted summary of activity within the developmental approach. Many of us forget that Freud, Hall, McDougall, Baldwin, and other psychologists at the turn of this century attempted to incorporate age-related changes in behavior and intellectual functioning in their explanations of human psychology.

With respect to intellectual functioning, Baldwin (1915), for example, advanced a broad description of intellectual development which, among other things, specified three stages in young children's interpretations of reality. In particular, he described a *pre-logical*, a *logical*, and a *hyperlogical* stage of intellectual awareness. Infants in the prelogical stage were described as encountering difficulty in distinguishing between their subjective and objective experiences. According to Baldwin, their understanding of the world was based upon a "meaning of immediate presence and intuition." Baldwin described his own infant's failure to recognize her bottle when it was presented to her nipple first, and speculated that this was because she had not yet constructed an overall *scheme* of the bottle that would enable her to identify it in different orientations.

Somewhat older infants, in the logical stage, have obtained a strategy for orienting themselves to stimuli like bottles both because they possess an increased capacity for focused attention and because they have achieved a higher degree of what he called "cortical coordination." Baldwin felt that this strategy represented both an awareness of the bottle's characteristics and an awareness of the orienting response.

Young children in Baldwin's hyperlogical stage possess imaginative schemata that are used to unite thoughts with logic. The two-year-old's schemata take

> the lead toward projecting its schematic and tentative readings forward with various shadings, semblances, and probabilities . . . Reality becomes embodied in all sorts of "as-if" constructions (p. 29).

This brief sketch of a small portion of Baldwin's theory is meant to underscore the fact that psychologists early in this century were, in fact, attempting to describe and explain age-related changes in intellectual development. Unfortunately, although Baldwin's theory was compatible with the best available behavioral data about children at the time, there simply was not much available. Thus it should not be surprising that, for much of the first half of this century, the primary efforts of psychologists interested in developmental phenomena were directed toward simply collecting information about children's behavior. Indeed, during this period a considerable effort was made to describe children's growth and development in normative-descriptive terms.

In the late 1930s and early 1940s, interest in practical concerns of childrearing and education within a normative framework began to shift toward interest in measurement of variables within an experimental framework. As Bronfenbrenner

(1963) put it, "the gathering of data for data's sake seemed to have lost its flavor" (p. 527). Although many psychologists, particularly clinical and educational ones, continued to be concerned about childrearing and education, more and more often this concern was expressed in terms of children's *learning* rather than their development. As Cronbach (1965) observed, during the 1940s and early 1950s:

> commerce between academic and educational psychology was cut off by a tacit embargo. Persons concerned with education found no nourishment . . . in the studies which began to dominate experimental psychology (p. 110).

Indeed, developmental psychology became less problem-oriented and more variable-oriented as researchers began asking questions about how experimental variables might affect children's learning, rather than simply how long it took them to learn a particular task. Attention was increasingly focused upon learning in general, not just children's learning per se. As Munn (1954) wrote,

> So far as anything fundamentally new concerning the learning process, the investigations on learning in children have failed. One possible reason for this is that such investigations have from the first been patterned too much after the lines of earlier research with adults and animals. A more likely reason, however, is that the phenomenon of learning is fundamentally the same whether in animal, child or adult (p. 449).

Toward the end of the 1950s, the attention of psychologists underwent still another shift, this time away from learning and behavioral changes and toward such topics as attention, language development, and problem solving. Interest in internal complex structures and processes was stimulated by Chomsky's (1957) alternative to Skinner's (1957) behavioral analysis of language. Bruner's (1964) ideas about education and the structure of knowledge; Newell, Simon and Shaw's (1958) description of mind in terms of a computer analogy; and Pribram's (1964) interest in the neurophysiology of mind. All these ideas served to increase the general receptivity by psychologists to theories that attempted to specify mental structures and the processes by which they developed. This receptivity was provided with a further stimulus with Hunt's (1961) efforts to relate Piaget's theory to existing maturational and behavioral theories, and with Flavell's (1963) intense summary of Piaget's theory.

Between 1964 and 1974, the dominant focus of psychological interest centered upon various "mentalistic" theories and, more specifically, upon studies of different aspects of intellectual development. By the mid 1970s, however, interest in trying to accelerate intellectual growth by training and other forms of intervention began to give way to an examination of a much broader range of age-related changes across the whole lifespan (Wohlwill, 1973). During the late 1970s and early 1980s, a number of alternatives, extensions, or variations of Piaget's theory began to appear, and psychologists undertook studies of metacognition, attribution, and various aspects of social cognition. Correspondingly, studies of spatial development shifted from a preoccupation with object-centered relationships to large-scale relationships.

These historical shifts of interest by developmentalists can be broken down into three phases. In the first phase (1900–1960), psychologists were primarily interested in observing and recording information about a wide range of children's normative behavior. In the second phase (1960–1974), attention was generally focused upon complex internal processes of intelligence. And in the most recent phase (1974–present), interest has broadened to include a concern about the competencies and strategies that people of all ages use in different spatial environments.

As in the previous chapter, a sampling of developmental models of psychological space will be described in each of these three phases. It should be noted, however, that these developmental models differ from those of the psychometric or experimental approach to the extent to which they can be depicted or represented graphically in figures. Where it is comparatively easy to depict the factor structures of psychometric models and the flow charts of experimental models, it is difficult to represent the complexity of such developmental models as Cassirer's, Werner's, or Olson and Bialystok's.

This chapter also differs from the psychometric or the experimental chapter in that it contains a brief description of three research topics as they evolved and changed across the three phases. These topical summaries are intended as illustration of the changes in emphasis by developmental psychologists over the three phases.

Phase 1: 1900–1960

For the first sixty years of this century, the primary activity of psychologists interested in development was the collection and interpretation of normative data about children's growth and behavior. Gesell's maturational model is a good example of this preoccupation, especially as he also proposed some interesting ideas about vision and spatial development. As interest shifted gradually toward intellectual development, Cassirer and Werner both advanced models that had more to say about spatial development per se.

In addition, interest in spatial development was expressed by research into three particular topics: children's drawings, their difficulty with left-right identification, and their gradual understanding of large-scale or geographic space. As concern with these topics began well before the turn of the century, we shall begin our discussion of this first phase by examining them before going into descriptions of the models.

Research on children's drawing has indeed had a long history. Barnes (1893), for example, described children's drawing responses to a verbally-presented story, and reported a definite sequence of representational development that corresponded to increasing age. Maitland (1895) observed that young children had a pronounced preference for drawing the human figure. In 1908, Stern argued that children characteristically experience themselves at the center of space in their drawings, and only gradually come to subordinate themselves to their surrounding environment in their drawings. By 1921, Burt had found evidence for seven stages of children's drawings. Young children, he observed, began by scribbling at two or three years of age, achieve a line stage at about four years, enter a stage of descriptive realism between ages five and six, attain a realism stage between seven and nine, a stage that Burt called "visual realism" between ten and eleven, a repression stage between eleven and fourteen, and finally a stage of artistic realism which, Burt noted, is experienced by only a few of us.

As Piaget cautions, to study the development of a child's idea of space on the basis of drawing alone would be an extremely hazardous venture.

> But assuming that this type of analysis is checked by other methods, and especially if one restricts oneself to the general features of drawings based upon simple everyday shapes, there can be no doubt that drawing does constitute a certain type of representation. "Pictorial space" is one of the types of representational space, and Brunschvicq (1922) went so far as to account for the origin of geometry on the basis of the technique of drawing (Piaget & Inhelder, 1956, p. 46).

Piaget is quoted here because he subsequently expanded upon the three stages of children's drawings that Luquet had earlier characterized as synthetic incapacity, intellectual realism, and visual realism, respectively. For Luquet, synthetic incapacity was characterized by a representation of space that neglects proportions, distance, and three-dimensionality. Intellectual realism is achieved when a child concentrates for a considerable period of time upon one type of drawing, and draws not what he or she actually sees but "everything that is there" (Luquet, 1927, p. 224). And visual realism is achieved when, at around eight or nine years of age, the child endeavors to take into account perspective, proportion, and distance all at once.

Burt's and Luquet's analyses were fairly typical for their time. From a clinical viewpoint, however, efforts to interpret children's drawings during the 1920s and 1930s usually reflected one of two other viewpoints. On the one hand, empiricists like Goodenough (1926) held that the analysis of children's drawings required hypothesizing about such higher mental processes as discriminations, associations, and the generalization of details and relations. According to Goodenough, these processes reside in concepts which children learn to manipulate. As she observed,

> It seems evident, then, that an explanation of the psychological functions which underlie spontaneous drawings of little children must go beyond the fields of simple visual imagery and hand-eye coordination and take into account higher thought processes. It has been said that the ability to precede any real attempt to represent objects in pictures, an ability which obviously precedes any real attempt to represent objects by means of pictures, is dependent upon the ability to form associations by the similarity of certain elements which are common to both picture and to the object, in spite of the dissimilarity of other elements. . . . In order to represent objects by means of pictures there must be, however, a conscious analysis of the process, of the intermediate steps by means of which the desired result is obtained. It is necessary to select from out of the total impression those elements or features which appear characteristic or essential. This analysis must be

followed or accompanied by observations of relationships. The relationships observed are of two kinds, qualitative and spatial. The former determine the proportion; the latter the position of the various parts of the drawing with reference to each other (p. 73).

If, as Harris points outs, children's drawings depend primarily upon concepts of objects rather than upon immediate visual images, then it is possible to understand why they increase in complexity while retaining a quality of "wholeness" as the children mature, and why children's developmental adaptations in drawings do not remain fixed from the time of their initial appearance (Harris, 1963, p. 194). With respect to this latter phenomenon, Goodenough observed that

the frequency with which any given characteristic tends to appear is a function of the extent to which it has become integrated into the developing concept, and a measure of weight should be given it as an index of concept development (p. 75).

Thus, when young children begin to add arms to their drawings of humans, they do not do so invariably; rather, as their concept of the human body develops, the arms tend to become an essential part of the body and are more consistently drawn. For Goodenough, children's drawings of objects are based upon their concept of those objects, which in turn are direct reflections of their experience with them. With age and experience the number of aspects of objects increases, and the relationships among them are better understood and are included in drawings.

On the other hand, psychologists who subscribed to the Gestalt view regarded children's drawings differently. Bender (1938), for example, held that they represented an expression of implicit neurological patterns immediately given or evident in the mind. As Harris said,

The Gestaltists held that perceptual processes, controlled by hypothesized neural actions in the brain, cause all stimulus situations to be experienced as "patterned" into figure-ground relationships. The responding organism is thus selectively patterned by neural activity, and may be subject to Gestalt principles (Harris, 1963, p. 175).

Bender (1938) administered nine of Wertheimer's (1923) geometric forms to children of different ages and to persons with various mental disorders, and evaluated the drawings of these forms according to the extent to which they embodied the features of the gestalt. She found a progressive improvement in drawing from a scribble response at age 3 to an increasingly improved representation of Gestalt up to about age 11. After age 11, improvement was essentially one of perfecting details such as sizes of shapes, distance, proportion, etc.

Both Goodenough's Draw-a-Man Test and Bender's Visual-Gestalt test are used by clinicians and researchers today as measures of intelligence or dysfunction. Although derived from quite different backgrounds, both tests have been faulted for the motor activity that they require, and for the fact that drawings are not simple translations of image representation since only two dimensions can be represented in the picture plane (Arnheim, 1969, p. 165). Moreover, as van der Horst (1950) points out, as children's lives become richer, more complicated, and more abstract, their drawings become progressively less adequate means for expressing concepts and feeling as other, more adequate modes of expression become available (cited in Harris, 1963, p. 182).

A second example of the descriptive, normative data collected during this first phase concerns young children's difficulties with left-right identification. Again, these difficulties were studied before the turn of this century. Badal (1888) advanced the hypothesis that left-right confusion "was due to a profound alteration of the sense of space." Others regarded this difficulty as "an impairment of the sense of direction in space," or as an "incapacity to apprehend figure-ground relationships" (Benton, 1959, p. 5). Gordon (1923) noted that children who correctly identified their own lateral body parts often encountered difficulty when asked to name the body parts of persons facing them. Head (1920; 1923) suggested that some degree of impairment in left-right identification was associated with aphasic disorders. Benton (1959) observed that children who exhibited systematic reversals in left-right identification were often retarded in their language development.

Researchers interested in children's left-right identification confusions were not long in seeing a relationship between difficulties with body-part identification and reversals of letters or words. Orton (1925) and Monroe (1932), for example, both noted that a characteristic of children with reading disabilities was their tendency to reverse or invert letters or letter sequences. Fildes (1923) stated that the tendency to confuse letter or word position or orientation was part of a larger tendency to recognize forms without apparent heed to the position that they occupied in space. As shall be seen, interest in left-right identification and in letter or word reversals has persisted through all three phases of developmental history.

The third major concern of the descriptive data collected during this initial phase concerned children's apparent difficulties with large-scale representation and their lack of understanding about cardinal directions (i.e., north, south, east, west). Many of the early studies were undertaken by geographers interested in children's efforts to draw maps of both familiar and less familiar terrains (Dodge & Kirchway, 1913; Rushdoony, 1968). With respect to adults' large-scale-representations, Trowbridge (1913) administered an "imaginary map" task, which consisted of subjects marking on a circular piece of paper the directions from the center to several distant places (e.g., North Pole, Panama, etc.). Adults employed two reference systems when performing this task: a *domi-centered* system, with home as the fixed object and "all changes of position at any moment referred to definite distances and angles, forming a simple trigonometric figure which gave direction to the home"; and an *egocentric* system, with subjects using reference points on a horizon, and imagining lines from these points intersecting at the body or ego. Trowbridge noted that some adult subjects used both systems to determine correct directions.

Freeman (1916) described a person visiting an unfamiliar city as constructing a system of orientation by unconsciously applying the cardinal directions to various landmarks or features in the city. He thought that in our first acquaintance with the different parts of a place, we adjust each by itself with relations to the cardinal points of the compass and we think of each part thereafter, without being fully away of it as related to these cardinal points (p. 167).

As for children, he felt that younger ones orient themselves only in terms of their own bodies, while older ones orient themselves with reference to the position of some fixed object or direction. He wrote that "as children mature they achieve a detached view of a region as though it were seen from a distance" (p. 165).

Almost twenty years later, Woodring (1938) found that differences in sense of direction could serve to divide children in a third-grade class into three groups. The first group were those who could point "north" despite blindfolding or other distractions; the second group pointed north after blindfolding only after deliberating plotting the number of left and right turns; and the third group—by far the largest—were unable to locate north accurately under any condition. When Woodring brought the first two groups together, he found that the children in the "intuitively-oriented" first group had difficulty communicating about map directions with those in the "reference-oriented" group. However, the children had little difficulty understanding each other when they worked on a mapping task within their own groups. Woodring's findings suggest that many factors may affect children's development of reference systems and awareness of directional orientation.

Children's drawings, their difficulty with left-right identification, and their development of large-scale systems of reference are only three examples of the many "spatial" topics that caught the early interest of geographers and psychologists. Although a large proportion of studies undertaken during this first phase was atheoretical (White, 1970, p. 662), it would be a mistake to assume that developmental theories were not being advanced. Some theories, like Gesell's (1940; 1949), were essentially just descriptions of changes of behavior with increasing age, but others, like Cassirer's (1944; 1957) and Werner's (1948) were broadly conceived accounts of mental development, with emphases upon changes in the structure and forms of thought. As some of the issues addressed by these theorists have affected the way in which others have thought about spatial development, their ideas require a brief summary.

Gesell's Model

In his many years as a physician, psychologist, educator, and writer, Gesell studied the importance of genetic and biological influences upon children's development. He maintained that maturation is controlled and determined by endogenous factors, which establish the direction and pattern of development. Based upon an enormous body of observations, Gesell believed that children's development progresses through an orderly sequence of behaviors (crawling, standing, walking, etc.) which, although its rate might be affected by accident, illness, or malnutrition, nevertheless continues throughout childhood. He also believed that "the child comes by mind as he comes by his body, through the organizing principles of growth."

Gesell's model consists of a set of principles that he derived from many years of observation. One such principle, that of *reciprocal interweaving*, was based on the observation that many human actions appear to have a dualistic quality. In the development of walking, for example, the body must alternatively bend and extend the leg muscles in a kind of neuromuscular coordination and integration over time. The back-and-forth quality of muscle dominance suggested the metaphor of weaving; Gesell believed that through such complementary processes, opposing sets of forces dominate at different times during a developmental cycle.

Gesell applied this principle of reciprocal interweaving to the study of visual and spatial development. In particular, he held that throughout childhood the child engages in

> multiplying and in shifting his planes of regard and in organizing his ability to manipulate space effectively on either side of each and every plane. In number, these planes soon approach infinity. In position, permutation, dimension, and configuration, the planes are also variously infinite. The task of development is to bring opposites into effective counterpoise. The development of vision is permeated with numerous bipolar opposites and paired alternatives: near/far; center/peripheral, monocular/binocular, abduction/adduction; vertical/horizontal; incoming/outgoing fusion; clockwise/counterclockwise; flexion/extension; skeletal/visceral; unilateral/bilateral and so on. The develomental mechanism which resolves counteracting forces in the process we describe as reciprocal interweaving (Gesell, 1949, pp. 159–160).

Moreover, he insisted that

> Newtonian physics and geometric optics cannot describe the relativities of the space world of the growing child. This world is his own private possession. He carves and constructs it out of a tenacious enveloping cosmos. He appropriates it by positive acts, reflexive, subconscious, autonomic, wished and willed. The space world thus becomes part of him. To no small degree, he is it. A description of his personality might well begin with the characteristics of this private space world of his. . . . The makeup changes with the developmental transformations of his total action system (ibid., p. 156).

For Gesell, vision does not automatically inform one about one's location because, at every stage of maturation, the visual system itself is undergoing changes. These changes, in turn, serve to re-orient the ever-changing individual.

Critics of Gesell claim, among other things, that his huge collection of age-related norms of behavior provides little information about how much variation in behavior might be expected for any given age. They also note that many of his age-related norms for children are outdated. However, it is important to remember Gesell's distinction between a visual and a spatial world; his descriptions of the bipolarities and paired alternatives that characterize visual development; and his emphasis upon action systems and self-regulatory processes.

Whereas Gesell observed and described children from a medical and psychological point of view, Cassirer (1944; 1957) was interested in the nature and evolution of thought from a philosophical viewpoint. A follower of Kant, Cassirer held that space and time provide humans with a framework for experience. Based on this psychological view, Cassirer undertook to explore some psychological aspects which in Kant's theory had been distinguished from the philosophical line of reasoning and thereby established as an empirical inquiry in its own right. It should be added, however, that Cassier's contributions to the problem of knowledge by far exceed what can be sketched here.

Within Cassier's framework, the emergent genetic function is not maturation per se, but the child's intention to know and its progressive tendency to construct knowledge about itself and about its world. For Cassirer, the key to both the differences between cultures and the differences between humans and animals was the *symbol*, the abstract designator of meaning. Humans differ among themselves and from animals in their capacity to construct, use, and interpret symbols through abstract thought. Since the construction of knowledge occurs within the experience of space and time, Cassirer believed that knowledge is a more important reason for psychological change than any other single factor. His extensive writings about the development of spatial knowledge had considerable influence upon Werner's and Piaget's theories.

Cassirer's Model

According to Cassirer, spatial knowledge developed in terms of three fundamentally different types of experience: *action space*, *perceptual space*, and *conceptual space*. At the lowest level, animals and infants experience an action space, which is directly tied to efforts at adapting to changing environments. As he observed,

> We cannot assume that animals, when performing these complicated reactions, are guided by ideational processes. On the contrary, they seem to be led by bodily impulses of a special kind: they have no mental picture or idea of space; no prospectus of spatial relations (Cassirer, 1944, p. 43).

For more complex animals and for young children, he described a second level of spatial experience; namely, the integration of sensory experiences in what he called perceptual space. Finally, he argued that action and perceptual space break down in the stream of experience and are replaced by an organized conceptual space. The latter becomes manifest when a person acquires a symbolic system that permits "that from the totality of experience a factor not only to be detached by abstraction but at the same time taken as representative of the whole" (Cassirer, 1957, p. 114). Cassirer insists that humans alone are capable of comprehending and representing the idea of abstract space, a conceptual or symbolic system used to extract invariance from the flux of our experience.

Like Werner, Cassirer was interested in the comparative analysis of mental development. He suggested a number of parallels between human and animal intelligence, between normal and pathological changes, and between mythical and contemporary conceptions of space. A persistent theme throughout his work is the distinction between our concrete acquaintance with, and our abstract knowledge of, space.

> The native is perfectly acquainted with the course of a river, but this acquaintance is very far from what we may call knowledge in an abstract, a theoretical sense. Acquaintance

means only presentation; knowledge includes and presupposes representation. The representation of an object is quite a different act from the mere handling of the object. The latter demands nothing but a definite series of actions, of bodily movements coordinated with each other or following each other. . . . But the representation of space and space relations means much more. . . . We must have a general conception of the object, and regard it from different angles in order to find its relations to other objects. We must locate it and determine its position in a general system (Cassirer, 1944, p. 46).

Cassirer's distinction between perceptual and conceptual space differs from Piaget's in that he attributes the development of knowledge to culture, while Piaget attributes it to the internalization of a child's actions (Olson, 1970, p. 17). In this regard, Cassirer (1957) anticipated anthropological findings that individuals living in non-technical societies often fail to achieve an abstract knowledge of space (e.g., Greenfield, 1976), not because they are incapable of it, but because neither their culture nor their language values the kinds of symbolic systems required for such representation.

Werner's Model

As Werner (1948) was teaching in the department of psychology the University of Hamburg at the same time that Cassirer was head of the department of philosophy there, it is not surprising that Werner's ideas about comparative analysis and development were influenced by Cassirer. While at Hamburg, Werner was also strongly influenced by Gestalt psychology and its argument that when we perceive things, we perceive whole forms (gestalts), which cannot be analyzed into separate parts. In his *Comparative Psychology of Mental Development*, Werner (1948) tried to show how the concept of development could be used to understand the development of mental life as a whole, including children, animals, psychotics, or peoples from different cultures. To bolster his arguments, he drew heavily upon anthropology, aesthetics, and embryology to create a broadly conceived (and often very complex) theory. For almost thirty years after coming to the United States in 1933, Werner revised and refined his theory of development in general, and his sensori-tonic model of spatial perception in particular.

It is important to remember, when discussing Werner's theory (Fig. 4.1), that it represented an effort to combine two interests: an organismic interest in the formal, structural, or organizational aspects of behavioral events from a comparative point of view, and a developmental interest in the formal changes that occur in the organism as it grows and develops. With regard to his organismic interest, Werner assumed that development is only one aspect among many transformations that affect an organism; other aspects include comparisons between normal/drugged individuals, normal/pathological individuals, etc. He believed that by focusing on the characteristic structure of a living system it was possible to analyze differentiated parts within the whole system for structural aspects, and mean-ends relations for dynamic or functional aspects.

With regard to his developmental interest, Werner and Wapner (1957) argued that development refers to more than changes over time (i.e., we may grow older

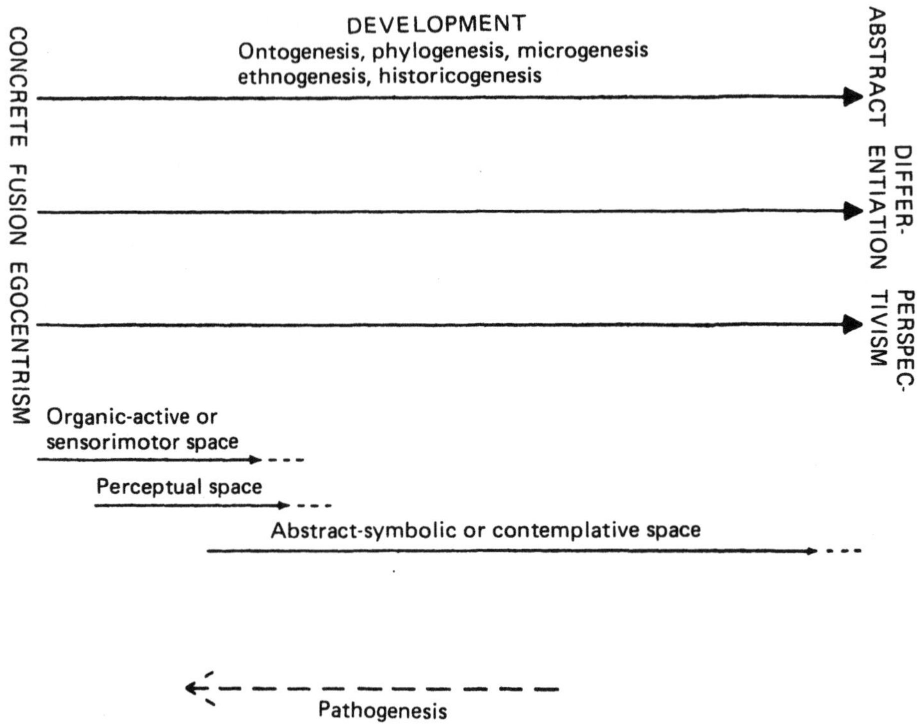

FIGURE 4.1. Levels of spatial experience recognized by Cassirer and Werner. (Reproduced with permission from Hart, R.A. & Moore, G.T. Development of spatial cognition: a review. In: R.M. Downs & D. Stea [1973]. *Image and environment*. Chicago, Aldine.)

without developing) or in size (we may also grow fatter without developing). Instead, he insisted that it refers to changes in structures as they proceed from a relative state of globality and lack of differentiation to a state of increasing differentiation, articulation, and hierarchical integration (Wapner & Werner, 1957, p. 126). From this *orthogenetic* principle, Werner identified three general stages of functioning: the sensorimotor, the perceptual, and the conceptual, with each successive stage reflecting an increased degree of part-whole articulation and means-end differentiation (Kagan & Kogan, 1970, p. 1284). At the sensorimotor stage, children's cognitive functioning consists of direct, motoric commerce with the world of objects; perception is global and whole qualities are dominant. The space of action or motoric commerce remains tied to specific activities. As Langer (1969) observed, children in this stage are likely to become disoriented if taken to a familiar place by an entrance they have previously used only as an exit (p. 762).

At the perceptual level, children's cognitive functioning is no longer limited to the immediate, although their symbolic operations are constrained to dealing with material objects. Their perceptions become more selective and are directed toward parts of configurations. By the age of 6 or 7 years, they may be able to identify their own left/right body parts, although not those of another person facing them. Moreover, they tend to mirror-image the movements of others when asked to imitate. Toward the end of this stage, "the construction of spatial coordinates begin to be detached from the children's actions. They can identify left and right on others, but, even at 8 years, they still have difficulty imitating other persons facing them" (Wapner & Cirillo, 1968).

Differentiation between subject and object is most advanced in the conceptual stage. "[The child] begins to take account of and to differentiate between quality (intension) and quality (extension) of the elements in a spatial configuration when he classifies them into wholes and parts. However, he still does not adequately integrate the intensive and extensive characteristics of class membership. Consequently, he still vacillates in his usage of all or some" (Langer, 1969, p. 143). For Werner, in short, development at any moment in the lifespan is expressed in terms of the degree to which an organism is organized, as well as the extent to which the separate components of this organization are in communication with each other while functioning independently.

Within the three stages of cognitive functioning, Werner treated the progressive development of an action-in-space to a perception-of-space to a conceptions-about-space as a function of increasing differentiation, articulation, and re-integration between children and their environment (Hart & Moore, 1973). Unlike Cassirer's progression, however, Werner's concern was with the interrelationships among sensory modalities, and with such problems as the contribution of muscular states to spatial perception. In particular, he assumed that "perception is a reflection of the relation between proximal stimulation and ongoing organismic states. Proximal stimulation was defined as stimulation of the sensory surfaces which issues from physical objects; organismic states were defined as the total ongoing state of the organism as it was affected by its past history, present internal stimulation, and stimulation from other sources" (Wapner, 1969, p. 8). Given the reciprocal relation between children and their environment, Werner argued that there can be no perception of objects without a bodily frame of reference, and no perception of the body without an environmental frame of reference. Put differently, he said that an "analysis of spatiality could not be complete without considering the interconnections between localization of the body and localization of the object under similar conditions" (Wapner, Cirillo, & Baker, 1971). Accordingly, Werner undertook studies that systematically varied body position in space under varying conditions of body tilt. He also studied such problems of perception as object localization, perception of size of body parts, perception of part-whole relations, etc. His efforts to study spatial perception in terms of proximal stimulation and organismic states were, in short, clearly a logical extension of his organismic-developmental approach.

It must be emphasized that this brief summary of Werner's theory and model are drastic abbreviations of his ideas. Both his comparative theory and his sensori-tonic model constitute a very rich source of potential research ideas and, although some critics complain that the abstractness of his ideas inhibits their translation into operational form, it is doubtful that the processes of development are themselves much simpler. In particular, Werner reminds us that our perception of space involves much more than the maturation of vision or the mental construction of relations among objects and ourselves. No comprehensive theory of psychological space can ignore the complex interrelationships between sensory modalities and the contribution of muscular states to the ways in which we perceive objects and the relations among them.

Although a large proportion of research with children during this first phase has been described as essentially atheoretical, we have encountered a variety of theoretical orientations, from Goodenough's associationism, to Bender and Werner's Gestalt rationale, to Gesell's emphasis on biological factors in maturation.

Phase 2: 1960–1974

Between 1960 and 1974, what had been a modest stream of developmental literature swelled to an enormous river of models and research findings, the largest proportion of which was devoted to intellectual development. This included a dramatic increase in develomental research on a variety of spatial topics. As with the first phase, we will begin with reference to ongoing research in children's drawing, their left-right identification confusion, and their abilities of large-scale representation.

With respect to children's drawing, Harris (1963) reviewed the literature and concluded that graphic development evolves through three broad stages. In the first stage, young children appear less interested in actually representing things in the visual world and more interested in simply producing effects with crayon, pencil, or paint. Harris felt that only gradually do their marks on paper achieve form and character. Somewhat older children attempt increasingly complex drawings of their surrounding world, drawings which Harris thought could be used as indices of their cognitive complexity. As children approach adolescence, they learn to incorporate such graphic conventions as design, balance, and arrangement in order to produce pleasing effects and to communicate conceptually with one another. Harris based his revision and extension of the Goodenough Draw-a-Man test upon this progression.

During these years, the issue was not so much the difference between perception and representation, but how children's ongoing perception affected their representation and vice versa. On the one hand, Gombrich (1960) wrote that art developed through the progressive employment of a vocabulary of forms, and what a child "saw" in nature depended upon the limits of his vocabulary and the system of schemata he brought to bear on a particular drawing task. Gombrich said that the schemata could be compared to a questionnaire that helps to select information from the visual world: only those aspects of information that are judged to be useful or relevant are registered in the schemata:

the correct portrait, like the useful map, is an end product on a long road through schema and correction. It is not a faithful record of visual experience, but the faithful construction of a relational model (p. 90).

As a result, children's copying of visual forms involves more than what is immediately perceived—it reflects their private experiences and their cultural history as well.

On the other hand, Arnheim (1969) wrote that every act of seeing is a visual judgment. That is, children's drawings are simpler, less complex, and less differentiated than adults' because children respond visually rather than conceptually to the surrounding world. Their drawings lack content not because they fail to see content, but because they lack the ability to draw what they see.

The incapacity exists because perception is not primarily the apprehension of shapes or colors. Rather, to perceive is mainly to receive the expressive or dynamic effect of shapes or colors . . . when an average person looks . . . at a human face, he sees its tenseness or plumpness or slyness or its harmony rather than the colors or shapes producing that effect. . . . The difference, then, is not primarily between perception and representation but between the perception of effect and the perception of form, the latter being needed for representation (p. 206).

Simplification in children's drawings does not necessarily prove that they are unable to grasp the structure or the pattern of what is being drawn. Arnheim points out that

one of the reasons for "incorrect" reproduction is that unless a person has received specific instruction in mechanically correct copying, he tends to look for the overall structure of the model rather than imitating it painstakingly, piece by piece. Gustav Jahoda describes this approach in his experiments with Nigerian teen-age boys, who were given some of the Goldstein-Scheerer tests. Instead of matching one block after the other successively, the boys would look at the model figure for a while, then concentrate upon reproducing it, without more than cursory further glances at the model (p. 198).

The distinction between perceiving and representing in drawing was earlier an issue about the developmental lag between children's perceiving and performing. Maccoby and Bee (1965), for example, found that children who had difficulty reproducing a design often had no difficulty identifying it when presented in a group of similar ones. They concluded that

to reproduce a figure, the subject must make use of more attributes of the model than are required for the perceptual discriminations of the same model from other figures (p. 375).

Olson (1970), however, questioned whether additional looking by children would necessarily make the attributes more available. He asked several hundred children to draw an oblique line or to construct a diagonal pattern of checkers on a checkerboard. Many encountered difficulty with such tasks, and Olson argued that when children perform such tasks it is the sequence of correct choices they remember and represent rather than the action schema per se. Moreover, he claimed that while perception was a process of differentiating alternatives or choices in the performance of tasks, intelligence was the mastery of the media in which preformatory actions took place. Thus, an activity such as drawing is only one means of representing the visual world; direct instruction could be another (p. 197). Direct instruction could provide children with guidance about action choices in the performance of tasks in different media. Like Gombrich, Olson thought that children's drawing involved more than what they immediately perceive; unlike Gombrich, he based his argument on performance of a specific problem situation.

With respect to the confusion about left-right identification, Kephart (1966), Cratty (1967) and others interpreted results from normative studies as providing support for a developmental progression of awareness of space from practical to subjective to objective. In particular, these studies appeared to have in common a progression that began with children's awareness of left-right relationships on their own bodies (although not labeled as such); evolved as named relationships in terms of specific body parts; were transferred as relationships on others in relation to self; and finally became manifest in terms of objects in relation to other objects (see Ilg & Ames, 1964; Cratty & Sams, 1968, for additional evidence). Perhaps not surprisingly, this general progression corresponded to similar progressions described earlier by Cassirer and Werner. Unlike these earlier descriptions, however, researchers now were interested in a variety of possible interventions to remediate or accelerate this progression.

With respect to large-scale representation and geographical orientation, the publication of Lynch's (1960) book attracted many psychologists because it opened up new ways of thinking about large-scale space generally and "way-finding" in particular. Lynch based his ideas upon interviews with residents of various cities whom he asked for verbal descriptions and for sketch maps of their cities. After analyzing these verbal responses and sketch maps, Lynch concluded that residents' images of their cities possessed five characteristics: paths, landmarks, edges, nodes, and districts. He observed that the maps emphasized paths or routes between places, with large portions of the cities left blank. Moreover, he noted that there was considerable similarity between the maps and people's composite verbal descriptions.

According to Lynch, residents drew their maps in one of five ways: outward from familiar paths or lines of movement, outward from a familiar landmark or place, inward from an enclosing geographical boundary (e.g., a river or highway), in a gridiron pattern, or in terms of adjacent regions (p. 86). He concluded that the structure of a city had a dramatic effect upon a resident's ability to describe or to draw it.

As for way-finding, Lynch wrote that

> . . . wayfinding is the original function of the environmental image, and the basis on which its emotional associations may have been founded. But the image is valuable not only in this immediate sense in which it acts as a map for a sense of direction or movement; in a broader sense it can serve as a frame of reference within which the individual can act:

> it is like a body of belief or a set of social customs: it is an organization of facts and possibilities (p. 125).

This characterization of an image serving as a cognitive map for large-scale representation appeared at a time when psychologists were becoming interested in problems of children's environmental representations at different ages.

In the same year that Lynch described resident's representations of cities, von Senden (1960) summarized his studies of the spatial frameworks used by blind subjects. Interestingly, there were several parallels between sighted residents' representations and blind subjects' spatial frameworks:

> The (blind) subject sets out from a perfectly definite point, resembling the center of a spider's web, whence he gains acquaintance with the routes that matter to him, up to the outer periphery that can be more or less gradually extended. Wherever he may be, he always remains mentally in conscious relation to his fixed starting point. . . . After a blind person has successfully represented a number of journeys as "route maps," he then seeks to relate a number of such routes, all starting from the same point, by means of lateral crossties, and so to create a network of relations, to encompass and to knit together the parts of the region commanded in this way (p. 286).

During the 1960s, several geographers and psychologists cooperatively investigated age-related differences in large-scale spatial representation. Examples of such studies include the High School Geography Project (Pattison, 1964), The Cognitive Study of Basic Geography (Eliot, 1962), and the Place Perception Research Project (Blaut & Stea, 1969).

With regard to the latter, more than fifty studies were undertaken with children. These included research on map linguistics, on the relationship of toy play to cognitive mapping, on cross-cultural studies of land-use preference, on the relation of map-learning to reading, and on children's mapping from aerial photographs. Among the findings was evidence that children learn about large-scale environments in a somewhat different way than they learn about other things, and that their environmental learning occurs much earlier than had been supposed. Children at three years of age, for example, were found to be able to

> solve all essential problems of mapping-rotation from a horizontal to an orthogonal view of the landscape, reduction of scale, and abstraction of semi-iconic signs—before they were exposed to maps (Stea & Blaut, 1973, p. 59).

Moreover, by five years of age, children were found to have formed a highly evolved cognitive map of their own large-scale environment. However, Liben's and Downs' (1986) recent work fundamentally questions these conclusions. Similarly, McDonald (1986) found no evidence that rotation of perspective is important to learning to read maps.

Another example of research activity initiated during this period was the investigation of children's representations of large-scale environments by Pick (1972) and his associates. Pick began by studying the frames of reference that children used to orient themselves in different spaces. By frames of reference, he meant a set of interrelated coordinate points (and perhaps direction) with respect to which other points are located. He noted that adults operate in terms of a large set of nested or partially-nested frames of reference, with respect to identifying the location or the position of objects. He pointed out that an object may be referred to oneself, to its location to other objects in a room, to its relation to the room as a container, or by specifying its geographical relationship to a landmark such as a street or a compass orientation. Pick began more than a decade of investigations by noting that while adults may employ some or all of these frames of reference according to circumstances, it is not clear whether children do likewise, or whether different combinations are more apt to be employed at different ages.

Ittelson (1973) further stimulated interest in large-scale spatial representation by pointing out that object-centered space differs from environment-centered space in at least four important ways. First, whereas object-centered space requires only an observer, environment-centered space requires a participant. Second, traditional studies of object-centered space are usually visual in nature, but participation in environment-centered space entails the processing of supplementing, conflicting, and distracting information through several sensory modalities simultaneously. Third, whereas object-centered space depends upon an awareness of centrally focused information, environment-centered space involves both an awareness of peripheral and central information

> both in the mechanical sense that the area behind one is no less part of the environment than that in front, and peripheral in the sense of being outside the focus of attention. Both meanings are important and raise questions concerning the process underlying the direction of attention (p. 14).

Finally, he noted that whereas object-centered space gains meaning from a given surrounding, environment-centered space is characterized by a host of meanings, which different actions and purposes call forth.

These distinctions represent a considerable advance over Lynch's (1960) earlier efforts to describe how residents of cities represent and sketch large-scale spaces. As will be seen, psychologists were now in a position to make the study of large-scale space a research focus in its own right.

Although it is tempting to elaborate further on these three research topics. We will now move on to the most central issue in this second phase: Piaget's theory of intelligence and his account of spatial representation. Bruner's and Vygotsky's ideas will also be mentioned, because Piaget's model was not the only explanation for the course of cognitive development to interest psychologists during this phase. We shall begin by discussing Piaget's work, and then refer to these alternative models.

Piaget's Model

It would be a mistake to assume that Piaget's ideas about intellectual growth remained so comparatively fixed throughout his academic career. Indeed, Piaget himself never considered his theory "complete," and pictured himself as its chief revisionist. According to Montangero, Piaget's theory underwent four major revisions, with each revision resulting in a somewhat different model. Between 1923 and 1932, for example, he studied children's language, moral reasoning, etc., and described development as a progression from a state of egocentrism to one where different viewpoints were coordinated and where rationships with others were based upon reciprocity. But during the 1930s, he conceived of development more in terms of adaptation, with interaction no longer emphasized between subject and subject but rather between subject and object. Piaget's adaptation model stressed the functional aspects of knowledge; namely, he argued that to know something meant that it must be incorporated into existing action systems, and to gain in knowledge required that these action systems be incorporated into more and more action schemata (Montangero, 1985, p. 25).

During the 1940s, Piaget turned from functional aspects of knowing to address the structural aspects of intellectual development:

> Mental structures were considered as schemes upon schemes, not static entities but transformations which permitted the subject to actively organize the data of reality. (ibid., p. 26).

It was during this decade that Piaget wrote about children's conceptualizations of time, number, volume, space, etc.

Finally, in the 1960s, Piaget attempted a new synthesis which encompassed some of the dichotomies in his theory—between structures and functions of development, between the subject and the object of knowledge, between biology and epistemology, between action and thought, etc. It was not until this time that his ideas were subjected to intense scrutiny and testing.

Montangero observes that Piaget's construct of egocentrism is an excellent example of the evolution in his thinking. In the 1920s, the notion of egocentrism was central to his explanation of prelogical thinking. In the 1930s the idea of egocentrism acquired a descriptive character and was used to define the limits of knowledge in the early stages of intellectual development:

> The use of this concept was justified because it could be related to the main explanatory concept of adaptation. In effect, egocentrism is related to the assimilation of reality to the subject's own actions. However, the construct . . . ceased to play an important part in the society of ideas of Piaget's theory (ibid., p. 30).

In the 1940s, Piaget referred to egocentrism as the inability to take a different viewpoint other than one's own. Wallon (1945) strongly criticized this construct and, because it was not theoretically necessary, Piaget ceased employing it by the late 1940s. As Montangero notes, researchers who employed or criticized the construct of egocentrism in the 1960s probably did not realize that, from the point of view of Piagetian theory, they were dealing with a "theoretical ghost."

As to Piaget's theory itself, it is important to remember that he "inherited" large portions of it from other theorists before him. As Case points out, while Piaget de-emphasized Baldwin's concepts of attentional power and cortical coordination, he incorporated practically all of Baldwin's other concepts into his theory of intellectual development:

> As will no doubt be apparent, Piaget charted the relationship between the form of children's internal activity and their apprehension of the external world in much greater detail than Baldwin had. However, in explaining the new data he had gathered, he preserved and explicated many of Baldwin's concepts, most notably: 1) the notion of schema, 2) the notion of a circular reaction, 3) the notions of assimilation and accommodation, 4) the notion of developmental stage, and 5) the notion of qualitatively distinct types of logic. In addition, Piaget introduced a number of new concepts, including 1) the notion of a logical structure, 2) the notion of an invariant whose construction is permitted by that structure, 3) the notion of cyclic reconstruction of structures at higher levels (vertical decalage), 4) the notion of equilibriation, and 5) the notion of reflective (as opposed to empirical) abstraction. The new result of all these changes was that the overall picture of the young child that emerged from his theory was considerably different from the one that had emerged from Baldwin's. Baldwin's predominate image of the young child had been a biological organism, whose development recapitulated that of its species. While Piaget preserved the biological metaphor, his predominant image of the young child was that of the young scientist building successively more sophisticated models of the world, by the application of successively more sophisticated logical structures (Case, 1985, p. 22).

As there exists a very large number of books, journal articles, and monographs written since 1960 to describe, interpret, or criticize Piaget's theory, there seems little point in attempting yet another summary here. However, it will be useful to sketch his model of spatial development, although given the constraints of this book such a summary will necessarily be brief. In particular, we will include mention of his major works on space, four major assumptions about spatial development, and some commentaries about his study of perspectivism.

Piaget set out his ideas about spatial development in three major works: *The Child's Conception of Space* (Piaget & Inhelder, 1956), *The Child's Conception of Geometry* (Piaget, Inhelder, & Szeminska, 1960), and *Mental Imagery in the Child* (Piaget & Inhelder, 1971). In these three books, a number of important distinctions were made—namely, between perceptual and representational thinking, between figurative and operative thinking, and between reproductive and anticipatory imagery. For Piaget, thought differs from perception in that it is both structurally and developmentally superior. It does not change with one's location or with the direction of one's attention. Although Piaget held that the early stages of children's thought are dominated by perceptions, he argued that these perceptions, unlike thought, are capable of only limited development (Piaget & Inhelder, 1956, p. 451). It is not surprising, therefore, that he insisted that the construction of a conceptual space, requiring intensive interaction with objects in space, is indeed a long developmental construction.

The distinction between *figurative thinking*, which is concerned with the perception of static patterns of figures and objects and the formation of static images, and *operative thinking*, which is concerned with the perception of patterns in the movement of objects and the ability to manipulate visual images, was also important to Piaget and Inhelder, who maintained that it is operative thinking that is vital to mature spatial thought.

The distinction between reproductive and anticipatory imagery concerns the development of the functional abstractness of imagery. At the outset, Piaget and Inhelder (1971) defined imagery as "internalized imitation," and argued that imagery is characterized by its motoricity. Reproductive images are further divided into static, moving, and transformational reproductive images; anticipatory images are further divided into images of movements and transformations. Anticipatory imagery develops along with concrete operations and may be dependent on it.

> two main periods of image development correspond to the pre-operational (before seven to eight years) and the operational levels. As we have already pointed out . . . the images of the first period remain essentially static and consequently unable to represent even the results of movements or transformations and afortiore unable to anticipate processes not known. But at about seven or eight years, a capacity for imaginal anticipation makes its first appearance, enabling the subject to reconstitute kinetic or transformational processes, and even forsee other simple sequences (p. 358).

These three distinctions are worth noting, not only because their classification in Table 4.1 sets out a program of possible research, but because they are important in the subsequent discussion of Piaget's model of spatial development.

Piaget's model (Fig. 4.2) can be summarized in terms of four general assumptions: actions define space, spatial knowledge evolves in four stages, imagery is an essential component in spatial representation, and spatial knowledge entails three kinds of content. Each of these assumptions will be elaborated upon briefly.

The first assumption is that children's representation of space arises from a gradual *internalization* and *coordination* of actions. As Sauvy and Sauvy (1974) put it, our actions define our spaces and, in turn, our spaces are defined by our

TABLE 4.1. Piaget's and Inhelder's Scheme for Imagery Research

Images	Immediate (I = foreimages) or Deferred (II)	Bearing on Product (P) or on Modification (M)
Reproductive (*R*)		
Static (*RS*)	*RS I* or *RS II*	
Kinetic (*RK*)	*RK I* or *RK II*	*RKP* or *RKM*
Transformation (*RT*)	*RT I* or *RT II*	*RTP* or *RTM*
Anticipation (*A*)		
Kinetic (*AK*)		*AKP* or *AKM*
Transformation (*AT*)		*ATP* or *ATM*

Each of the images thus classified may be in itself either gestural (imitation), mental, or graphic.

Reproduced with permission from Piaget, J. & Inhelder, B. (1971). *Mental imagery in the child*. New York, Basic Books, p. 6.

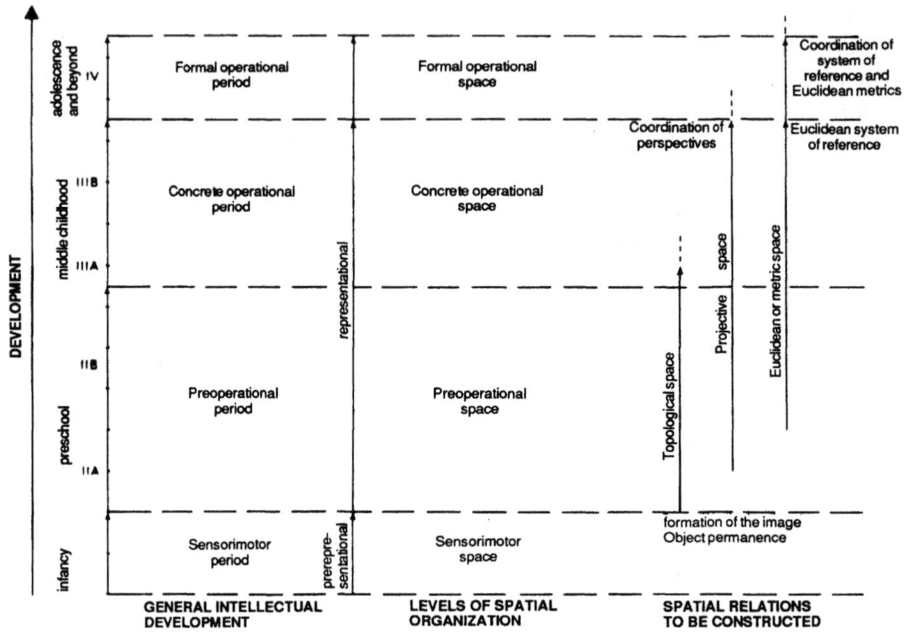

FIGURE 4.2. Piaget's developmental account of spatial cognition in relation to overall intellectual development. (Reproduced with permission from Hart, R.A. & Moore, G.T. [1971]. *Development of spatial cognition: a review.* Chicago, Environmental Research Group.)

activities. A parking lot, for example, is usually defined as a limited space for parking cars. But if this space is used to play a game of football when not used as a parking lot, it becomes a different space. Similarly, in the game of chess, the moves of the chess pieces, the pieces themselves, and the chessboard give a framework to the game by defining what can happen. To communicate about this framework and about the rules of the game we devise ways to communicate about the movement, orientation, direction, and distance of pieces. In the end,

> the word "space" has always implied some action that is possible. In the present-day world, the possibilities for action are so diverse and of such different kinds that no one definition of "space" will do (Sauvy & Sauvy, 1974, p. 12).

For Piaget, all knowledge of space is based upon actions or internalized actions. Perception is the knowledge of objects resulting from direct contact with them; representation involves the evocation of objects in their absence or, when it runs parallel to perception, in their presence.

The second assumption is that knowledge of space evolves through four levels or structures of organization: namely, sensorimotor, preoperational, concrete operational, and formal operational space. Each of these stages is characterized by the following characteristics: 1) they are qualitatively different from preceding ones; 2) they integrate the achievements or structures of all preceding ones; 3) they are both a consolidation of preceding stages and a preparation for the next; and 4) they are the manifestation of a constellation of behaviors at a similar level of achievement (Laurendeau & Pinard, 1970). Each stage thus consists of an organized totality of mutally dependent and reversible behavior sequences known as schemata.

The third assumption is that three classes of spatial relations form the content of spatial intelligence. The child's first spatial undertanding is topological, with projective and Euclidean content being understood after mastery of certain basic topological concepts. Each of these three classes of spatial content is characterized by different properties.

The child in the sensorimotor period is concerned primarily with topological relationships inherent in the figures or objects whose properties they express. Specifically, Piaget and Inhelder say that the child's object-space is dominated by such relationships as *proximity*, or the nearness of elements in the same perceptual field, *separation*, or the segregation of parts from wholes, *order*, or the relative positioning of neighboring but separated elements, *enclosure*, or the distinction between open and closed figures, and *continuity*, or the synthesis of the first four relationships in the appreciation of whole figures or objects. These topological relationships are regarded as independent of changes in apparent size or shape of the figures, and hence the child has difficulty trying to conserve such features as angles, straight lines, and distance if they change in orientation. By about two years of age, however, the child becomes aware of the difference between stationary and transitory objects, has an appreciation of object permanence, understands that the location of an object is independent of the path taken to reach it, realizes the difference between one's own and external movements, and recognizes the advantage of combining sensory modalities to obtain more information. As Piaget and Inhelder remark, it is a serious mistake to underestimate the significance of these understandings upon later intellectual achievements.

By about two years of age, Piaget and Inhelder say, children begin to show evidence of being able to represent the world to themselves. This evidence includes the fact that they can mime some experiences even if they are not yet able to verbalize them. It is also expressed in their practical knowledge about objects, and about the spaces in which both they and the objects exist and move. However, Piaget and Inhelder point out that this practical perceiving-and-doing kind of understanding is still far short of the achievement of symbolic-representational understanding of the same phenomena. Several more years of experience are required before the young child achieves a measure of objectivity in symbolized, imagined, and reflected-on space.

For most of the pre-school and elementary years, the child is described as being bound by concrete, perceptual experiences. The child is aware of topological relationships as concepts that the mind renews and recreates as correlates of activity at various times. Most children during this age-span are characterized as being unable to represent either their own or another's visual acts from a different viewpoint. But as various concrete operational skills are achieved in late middle childhood, spatial awareness gradually becomes disengaged from static imagery, and transforms itself into operations that allow the coordination of object arrangements and the abstractions of these arrangements in terms of multiple viewpoints. The child gradually understands that perspectives constitute a coherent intergrated system; that a given position vis-à-vis an object defines one and only one view of it; that right becomes left when seen from the opposite side; and other relevant facts.

The parallel achievement of projective and Euclidean space occurs when the child no longer views a figure or object in isolation, but considers it in relation to a *point of view*. In Piaget's and Inhelder's terms, the child achieves a frame of reference, which is fundamental to his subsequent ability to deal with the orientation, location, and movement of objects in space. As Flavell (1970) observed:

> what a child eventually needs to establish ... and does not first possess, is a picture of space as some kind of container made up of a network of sites or subspaces, which move from site to site, now occupying a given site, now leaving it unoccupied or empty (p. 1016).

Piaget and Inhelder (1956) characterize this frame of reference as a simultaneous organization of all positions in three-dimensions, with the axis within the reference system being objects or positions that remain invariant under an action or transformation within the system. (p. 377).

Whereas projective space includes the ability to respond to the intercoordination of objects separated in space, Euclidean space refers to metric criteria that one imposes upon physical spaces in order to communicate the relative positions of landmarks or to describe distances accurately. To explore the parallel achievement of projective and Euclidean space, Piaget and Inhelder had children construct small-scale models of large-scale environments in order to assess their understanding of the proportionality of scale, and their understanding of themselves as one object among a multiplicity of objects. Only after children had achieved the projective coordination of area were they able to deal with such problems as the distances between objects and their relative locations. As Mandler notes,

> metric properties add restrictions to the topological considerations. These properties begin to appear when considering projective space, in which objects are described in relation to each other through lines of regard or points of view. In projective space, linear order, and the distinction between straight and curved lines is thus made; in addition, rectilinearity remains invariant when the point of view changes. Euclidean concepts introduce the notion of coordinate axes, and include many of our most familiar notions about space, such as right angles and the metric concepts of distance (Mandler, 1984, p. 545).

The fourth assumption has already been alluded to—namely that for Piaget and Inhelder, imagery is a vehicle for representation, arising from the gradual internalization of deferred imitations. The young child begins by imitating others.

Over time this imitation schema is internalized and hence able to be deferred, and the child's responses become increasingly symbolic. The development of imagery is an important component in this model, because it is through imagery that the child achieves the capacity to perceive and to imagine the arrangement of objects from different viewpoints. More specifically, imagery is the capacity to make *transformations* on images, which enables one to master the representation of invariant properties in reversible displacements.

In their various works, Piaget and Inhelder investigated these four assumptions through a large number of experiments. One example is their demonstration of the achievement of projective space using the "three-mountain" task. In this experiment, they presented 100 children between the ages of 4 and 10 with a landscape model containing three prominent pasteboard mountains, and used three methods to assess the ability to perceive and imagine this landscape from different viewpoints. One method was to show a series of pictures taken at different locations around the landscape, and ask the children to match the picture that matched the view "seen" by a doll placed by the model. A second method was to ask them to select one picture and then to indicate where around the model the doll must have been located in order to have taken that picture. A third method was to present them with another set of pasteboard mountains, and ask them to reconstruct the mountains as seen in different pictures.

In their original study, Piaget and Inhelder found that subjects between five and six years of age tended to select a view comparable to their own viewpoint for all pictures. Children at this age were described as "centering" upon the relationship between themselves and the entire landscape model instead of the relationships between themselves and different features of the model:

> their representation were incomplete . . . and they responded to the task as if they assumed that to depict one relationship correctly, then all relationships would be depicted as a matter of course (Piaget & Inhelder, 1956, p. 323).

Nine- or ten-year-old children, by contrast, were more successful with these tasks. It was concluded that this was because they had achieved a cognitive system that enabled them to maintain the relative position of parts of the model, of parts relative to one another, and of the whole landscape relative to different points of view. For Piaget and Inhelder, what children read out of a perspectives task is determined by the operational system they were able to use: children will distort what they see in order to represent what they understand.

The three-mountain experiment has been described at some length here because it illustrates how Piaget made efforts to obtain empirical support for his ideas with ingenious studies. It is also important because, between 1960 and 1974, there were literally hundreds of efforts by others to replicate or to extend it, using a composite of Piaget's methods.

One difficulty with this later research has been the fact that different researchers have used a variety of landscapes, provided children with a variety of choice options, and have scored responses differently across studies. Without attempting a review of this literature, these variations deserve some comment.

With respect to differences in landscape stimuli, some investigators followed Piaget and Inhelder and presented children with rectangular or square landscapes (Dodwell, 1963; Neale, 1966; Kralovich, 1971), while others used circular boards (Pufall & Shaw, 1973; Barragy, 1970; Miller & Miller, 1969). Some researchers have used naturalistic mountains (Eliot, 1967; Ellis, 1972; Citrenbaum, 1975), and others familiar toy objects (Coie, Costanze & Farnill, 1973) or geometrical shapes (Laurendeau & Pinard, 1970). Some presented children with four pictures of the landscape (Laurendeau & Pinard, 1970), others eight or ten, either serially or at a single presentation. A number of researchers have also shown children pictures that were "impossible" views of the landscape (Fishbein, Lewis, & Keiffer, 1972; Houssaidas and Brown, 1967).

And finally, with respect to differences in scoring, some researchers would score responses as being either correct, partially correct, or incorrect (Dodwell, 1963; Kielgast, 1972), while others have used weighted scores to index accuracy (Flavell, 1972). Given all these variations, it should not be surprising that different research efforts failed to result in generalizable findings apart from age-related differences.

Generally speaking, Piaget's theory of intelligence and his developmental account of children's spatial representation became the central focus for many psychologists in the 1960s and 1970s. During this period, a number of variations on his ideas were also advanced, a few of which included spatial development in their focus. Two of these, Flavell's inference model and Fishbein's rule model, need mentioning before we consider Bruner's and Vygotsky's contributions.

Flavell's Model

Flavell (1972) extended Piaget's notions about perspectivism to account for children's role-taking behavior and the development of their tacit knowledge about different viewpoints. He argued that it was important to distinguish between the achieved level of development with respect to the child's understanding that different perspectives exist, and the level of achievement with respect being able to compute viewpoints in real time. To account for differences between children's knowledge about perspectives and their inferential skills in taking different perspectives, Flavell proposed a four stage model.

In Flavell's first stage, young children are at roughly the same level as Piaget's practical, sensorimotor stage in the sense that they are

> able to anticipate objects seen, but cannot anticipate or otherwise represent to themselves anyone else's seeing activity or experiences, per se (Flavell, 1968, p. 222).

By contrast, children in the second stage, while still unable to represent different perspectives, are able to realize that other people may see things differently. In studies of hidden objects with children between two and three years of age, Flavell, Omanson and Latham (1978) noted that most children could recognize that

> the other person could see the object when the screen was interposed between that person and the object. . . . They did not seem to mistake what they themselves did or did not see, even under task conditions designed to tempt that mistake (p. 370).

Children at stage three have come to understand that objects visible simultaneously to self and to other persons may have different appearances if seen from different distances or viewpoints.

> children at this level can grasp what for the younger child would presumably be a mind-boggling possibility: a subject-object difference in visual experience embedded in a subject-object similarity in visual experience (Flavell, 1971, p. 36).

Finally, children in the fourth stage are capable of reproducing a literal and precise characterization of different viewpoints, because they know that objects can have multiple appearances, and because they can make inferences about their perceptions.

Both Flavell's model and his supporting evidence indicate that children's achievement of the coordination of perspectives involves more than simply becoming aware of their own viewpoints; that young children's "seeing" differs from older children's in ways not accounted for in Piaget's model, and that children are capable of making inferences about objects earlier than Piaget believed.

Another alternative explanation for differences in perspectivism is Fishbein's rule-based model. Fishbein, Lewis and Keiffer (1972) argued that Piaget's experimental situation might be the reason why he failed to find patterns of inference among young children. Specifically, they point out that, with three-mountain tasks,

> We have always asked the child to identify the viewpoint of another human being or of a doll. The latter is in the realm of the hypothetical: "If a doll could see" or "If a doll could take a photograph." Dolls cannot see or take photographs, yet the child is asked to assume that they can. If young children have difficulty in dealing with the abstract or hypothetical, but can readily deal with the concrete . . . then methods which employ dolls instead of people should prove more difficult for these children (p. 32).

Fishbein's Model

Rather than describe a progression of levels of knowledge, from awareness of the existence of objects to an awareness of computational inferences about them, Fishbein and his associates suggested that the same phenomena could be explained in terms of the acquisition of a progression of *rules*. Specifically, they suggested three rules: 1) young children's rule use is basically *egocentric* or "you see what I see"; 2) somewhat older children use the *non-egocentric* rule or "if you aren't in my place, you don't see what I see"; and 3) still older children's rule use is basically *emphatic* or "if I were in your place, I would see what you see."

In research suggested by this model, Fishbein et al. found that as children exercised Rule 3 more often with increasing age, the use of Rule 2 decreased more rapidly than did the use of Rule 1. They concluded that "white middle-class children have clearly acquired rule 3 by age 3–5, and probably much earlier than that" (p. 31).

While Flavell's and Fishbein's models are examples of alternative explanations for some of Piaget's ideas about spatial perspectivism, two larger-scaled models of cognitive growth, by Bruner and Vygotsky, were also advanced during this period. We shall describe these two models here, as they both continue to affect thought about intellectual development in general.

Bruner's Model

Bruner (1964; 1966), like Flavell, emphasized a social context for the course of intellectual development and, like Fishbein, stressed the importance of children's acquisition of rules and rule systems. He differed from Piaget in that he characterized development in terms of the emergence of technologies that were acquired in "spurts as innovations were adopted."

> most of the innovations are transmitted to the child in some prototypic form by agents of the culture; ways of responding, ways of looking and imagining, and most important, ways of translating what one has encountered into language (Bruner, 1964, p. 13).

Unlike Flavell's emphasis upon role-taking, social interaction, and communication, Bruner emphasized a social context in the sense of the influence of language and culture upon the ways in which people think. Also, unlike Fishbein's gradual acquisition of rules for solving specific perspective problems, he emphasized the acquisition of rule-use through language as an important tool for adapting in a culture.

Although Bruner did not advance a model of spatial development, his ideas about modes of representation certainly included references to spatial dimensions. As he noted,

> If we are to benefit from contact with recurring regularities in the environment, we must represent them in some manner. To dismiss this problem as "mere memory" is to misunderstand it. For the most important thing about memory is not storage of past experiences but rather the retrieval of what is relevant in some useable form. This depends upon how past experience is coded and processed so that it may indeed be relevant and useable in the present when needed. The end product of such a system of coding and processing is what we make speak of as a representation, [the three modes of which are] enactive representation, iconic representation, and symbolic representation. Their appearance in the life of the child is in that order, each depending upon the previous one for its development, yet all remaining more or less intact throughout life (p. 2).

With respect to the enactive mode of representation, Bruner argued that actions became representational by virtue of their habitual patterning. As young children become more coordinated and skillful, they become freer of the need for immediate serial regulation while their behaviors are occurring. What was initially sequential becomes, with repetition, blended together and automatic.

Once a schema becomes abstracted from a particular act and becomes related to serial acts in a one-many relation, it can become the basis for action-free imagery. Then the young child's way of representing the world also becomes free of action. It is a slow procedure, this dissociation of a spatially-oriented schema from supporting action (Bruner, 1966, p. 19).

With respect to the iconic mode of representation, he held that young children are well on their way toward representing the world by the selective organization of their perceptions and images by the end of their first year of life.

it seems not unlikely that the properties one finds in visual perception . . . might appear in some exaggerated form in the child's imagery or his spatial schematization (p. 21).

Bruner argued, not unlike Galton before him, that iconic thinking, though useful in preserving past experience, was nonetheless dominated by rules of organization that rest upon spatio-temperal-qualitative properties of experience; rules that either aid or hinder children's performance on intellectual tasks.

Eventually, iconic representation gives way to symbolic representation. With this latter mode, children are able to represent things through language.

I tend to think of symbolic activity as some basic or primitive type that finds its first and fullest expression in language, then in tool-using and finally in the organization of experience (p. 44).

Symbolic representation, according to Bruner, is unique to humans. Where Piaget regarded thought in terms of internalized or deferred actions, Bruner emphasized the directive function of language on thought, and the role of culture and its interaction with development and learning in the child. Indeed, much of his more recent work has been concerned with the development of language, the relation between linguisitic and non-linguistic learning, and the development of competency in children.

Vygotsky's Model

Like Bruner, Vygotsky (1962; 1978) also emphasized the directive function of language on thought. But where Piaget and Bruner agreed that children construct a reality through actions and that language eventually becomes a vehicle for thought, Vygotsky believed that language and thought develop along different lines until children reach a point where thought becomes verbal, and speech rational. For Vygotsky, thought itself is internalized language, beginning as dialogue in a social setting, becoming a monologue, and finally becoming "inner speech."

Unlike Piaget or Bruner, Vygotsky distinguished between a *social* plane of psychological functioning, where children's learning and behavior is guided by the knowledge and instruction of others, and an *individual* plane, where their learning and behavior gradually come under their own cognitive control or regulation.

> Any function in the child's cultural development appears twice, or on two planes. First it appears on the social plane, and then on the psychological plane. First it appears between people . . . and then within the child. This is equally true with regard to voluntary attention, logical memory, the formation of concepts, and the development of volition (Vygotsky, 1978, p. 163).

Moreover, he described the difference between children's social and intellectual functioning as being a *zone of proximal development*. He defined this zone as

> the distance between the actual developmental level as determined by independent problem solving and the level of potential development as determined through problem solving under adult guidance or in collaboration with more capable peers (1978, p. 86).

The notion of a zone of proximal development is consistent with Vygotsky's belief that what children can do with the assistance of others may be more indicative of their mental development than what they can do alone, Piaget notwithstanding. Vygotsky's notion exists as an invitation to take a closer look at what is happening when adults intervene to help children solve problems.

An important component in Vygotsky's theory is the progressive refinement of children's ability to classify objects. His famous sorting test, adopted from Ach (1921), required children to form groupings of a set of blocks that varied in size, color, and shape. Vygotsky identified three stage in children's concept development: 1) very young children organized the blocks in *heaps* apparently randomly; 2) somewhat older children grouped them in *complexes* according to their apparent similarity or the fact that they appeared to share a common attribute; 3) older children sorted and grouped according to *logical rules* of inclusion ("all red, short blocks here, all white, tall blocks there"). According to Vygotsky (1962) children achieve true concepts when they are able to guide their mental operations verbally, using words "as a means of actively centering attention, abstracting certain traits, synthesizing them, and symbolizing them by a sign" (p. 81). He insisted that it is through language that a culture or society puts its own peculiar stamp on children's concept development, and through language that children share a treasure trove of social concepts.

Both Bruner and Vygotski's ideas about representation are important because they remind us how our language and culture critically influence how we come to see the world, represent it symbolically, and communicate about it with others. In contrast to Piaget's insistence that children construct their reality through action, both Bruner and Vygotski emphasized the role of language upon such construction. Both these theorists have also contributed to the study of the language of space, a topic that has been further investigated by such researchers as Bierwisch (1967), Teller (1969), Leech (1969), Clark (1971; 1973; 1977; 1978), Olson (1975), Talmy (1983), and others.

To conclude our summary of this second phase, we have seen how there were advances in theory and research on children's drawing, left-right identification, and development of large-scale spatial representations. We have described Piaget's theory of intelligence, and summarized the assumptions upon which he based his model of spatial development. To illustrate one research focus among many in Piaget's model, we have considered problems associated with researchers' efforts to replicate, vary, or extend his ideas about perspectivism. Finally, we have mentioned Bruner's and Vygotski's models of cognitive growth, because these provide alternative explanations for representation and address the role of instruction in fostering intellectual development. As an index to the richness of this second phase, perhaps the most succinct summary has been offered by Hart and Moore who noted that, by 1973, the currency of discourse among psychologists had grown to include

> five domains of parallel development: levels of organization (sensorimotor, preoperational, concrete operational, and formal operational), types of spatial relations (topological, projective, and Euclidean), modes of representation (enactive, iconic and symbolic), systems of reference (egocentric, fixed, and coordinated) and types of topographical representation (routes and surveys). Each of these developments in turn parallels the four periods of general intellectual development discovered by Piaget, and the orthogenetic principle and developmental progressions elucidated by Werner. Furthermore . . . there seems to be direct correspondences and functional and explanatory relations between the different domains of the development of spatial cognition (Hart & Moore, 1973, p. 288).

Phase 3: 1974–Present

Over the past dozen years, the study of psychological space from the developmental approach has become less the subject of broadly conceived models of spatial development, and more the focus of fragmented or isolated topics of interest. In this third phase, the developmental literature has been filled with studies of infant spatial competence, cross-cultural comparisons, training of spatial skills, spatial behavior of special populations, and the like. The topical nature of recent research has been reflected in the large number of books published within the last dozen years on infant spatial development (e.g., Bower, 1977; Butterworth, 1977), the physiology or neurophysiology of spatially-oriented behaviors (e.g., O'Keefe & Nadal, 1978; Potegal, 1982; Hein & Jeannerod, 1983), the emergence of sex-related differences in spatial behavior (e.g., Sherman, 1978), spatial learning strategies (e.g., Holly & Dansereau, 1984), spatial processing (e.g., Cliff, 1981; Isard, 1979), and large-scale spatial representation (e.g., Downs & Stea, 1977; Hart, 1979; Liben, Patterson & Newcombe, 1981; Pick & Acredolo, 1983; Cohen, 1985; Wellman, 1985). As these and other topical interests have become research domains in their own right, it becomes increasingly difficult to envision how such diverse strands will ever be re-united or integrated into a comprehensive explanation of spatial development throughout the lifespan.

The three such strands that have been described in the earlier phases will be brought forward again here. To begin, we will briefly discuss research interest in children's drawing, in terms of cross-cultural differences and the rules required by the act of drawing.

Since Hudson (1960) reported that Bantu children in primary grades encountered difficulty when asked to interpret drawings of a three-dimensional scene, the study of cross-cultural differences has expanded to include a variety of investigations into their graphic representation and interpretation (e.g., Miller, 1973; Kennedy, 1974; Bayraktar, 1985). When Mitchelmore (1976) undertook to summarize findings on the pictorial space of children and illiterates in developing countries, he listed five categories of findings:

1. Except in extremely isolated communities which have no experience of pictures, familar objects can be identified fairly accurately from a single-line drawing or photograph (Deregowski, 1968; Fonseca & Kearl, 1960; Holmes, 1963; Shaw, 1969; Spaulding, 1956).
2. Conventional signs such as those used to express movement are not well understood by children and illiterates in developing countries (Winter, 1963).
3. Cues used to represent depth in three-dimensional scenes are poorly understood (Dawson, 1967; Deregowski, 1968; Hudson, 1967; Vernon, 1969).
4. The frequency of correct interpretation increases with age, education, and urban influence (Dawson, 1973; Sinka & Shukla, 1973).
5. Depth cues of size and superposition are most easily interpreted, and perspective is the most difficult (Miller, 1973; Kennedy, 1974; Kilbride & Robbins, 1968).

Generally speaking, these findings support Gibson's (1969) earlier contention that intelligent viewing of a picture requires a knowledge of pictorial conventions on the part of the viewer; conventions that enable us to represent objects distinct and independent of their physical referents.

Mitchelmore also observed that children in many developing countries lag behind their more industrialized peers both in development of spatial abilities and in understanding of graphic representations of three dimensions. He pointed out, however, that most cross-cultural studies have employed correlational analyses, and that there are inevitably alternative explanations of difference. In particular, he noted that

> Eskimos may be superior to Africans in spatial ability because their environment requires more highly developed navigational skills, because they are brought up to be more independent as children and are less specialized as adults, because their language contains more spatial terms and their art is more intricate, or because their diet contains a greater proportion of protein (Mitchelmore, 1976, p. 166).

Mitchelmore's work is noteworthy not only for his summary of cross-cultural studies of pictorial perception, but because he devised two drawing tests for cross-cultural research. When he examined the responses of children between the ages of 7 and 15 from the United States, Jamaica, and England to his *Solid Representation Test*, he found evidence for five stages in the representation of a cylinder, a cube, a pyramid, and a cuboid. These five stages were:

1. an outline of the solid or the face viewed orthogonally;
2. several faces shown but not in correct relation to each other, often both visible and invisible faces shown, usually no depth depiction;

3A. only visible faces shown in correct relation to each other but with poor depth depiction;

3B. all appropriate faces distorted in an attempt to show depth, but not correctly; and

4. correct drawing using parallel or slightly convergent lines to represent parallel edges of the solid.

As he points out, these five stages correspond roughly to the four stages of meaningful drawing reported by Lowenfeld and Brittan (1964). Typical drawings at each stage of solid representation are presented in Figure 4.3. The reliability for his test was .93 across the three groups of children (Mitchelmore, 1976, p. 157).

Mitchelmore also developed a *Three-dimensional Drawing Test* with four drawing exercises requiring an understanding of parallels and perpendiculars. These included drawing telegraph poles alongside a road winding into the distance, water levels in four tilted bottles, lines on the faces of four blocks to make each block look as if it were made from cubes stuck together, and the hidden edges of four blocks. This test was administered to children of comparable educational level and urban background in Jamaica, the U.S., England, and Germany. He estimated the reliability of the test to be .85, and found evidence of considerable cross-cultural differences in performance (see Mitchelmore, 1978; 1980; 1982). Mitchelmore's work has been described in order to underscore the fact that there appears to be an important cross-cultural dimension to the study of children's graphic representation (e.g., see Hazen, 1983).

With regard to interest in rules of drawing, it is found that within industrial societies children may be more successful in drawing solid geometrical objects, but they nonetheless have difficulty before the age of 8 in drawing familiar objects from different viewpoints. In particular, Goodnow (1977) found that until children are about eight years of age they insist on drawing only the front view of a human face and the side view of a horse, even when asked to draw them from different orientations. Similarly, Hagen (1976; 1980) found children deficient in perceiving the spatial properties of pictures when viewed from oblique angles. According to Freeman and Cox (1985), Hagen concluded that

> there was no compelling reason to think that children chase through all four systems of projective depiction in pursuit of the elusive perspective mode (p. 3).

Freeman and Cox (1985) have also argued that children's knowledge of pictorial conventions differs from their knowledge of the rules for graphically representing objects. Indeed, Freeman (1980) has wondered whether it is possible to gain a clearer understanding of children's knowledge of drawing rules from their actual drawings.

From a somewhat different vantage-point, van Sommer (1984) has argued that drawing, like talking, is a complex skill that requires a considerable amount

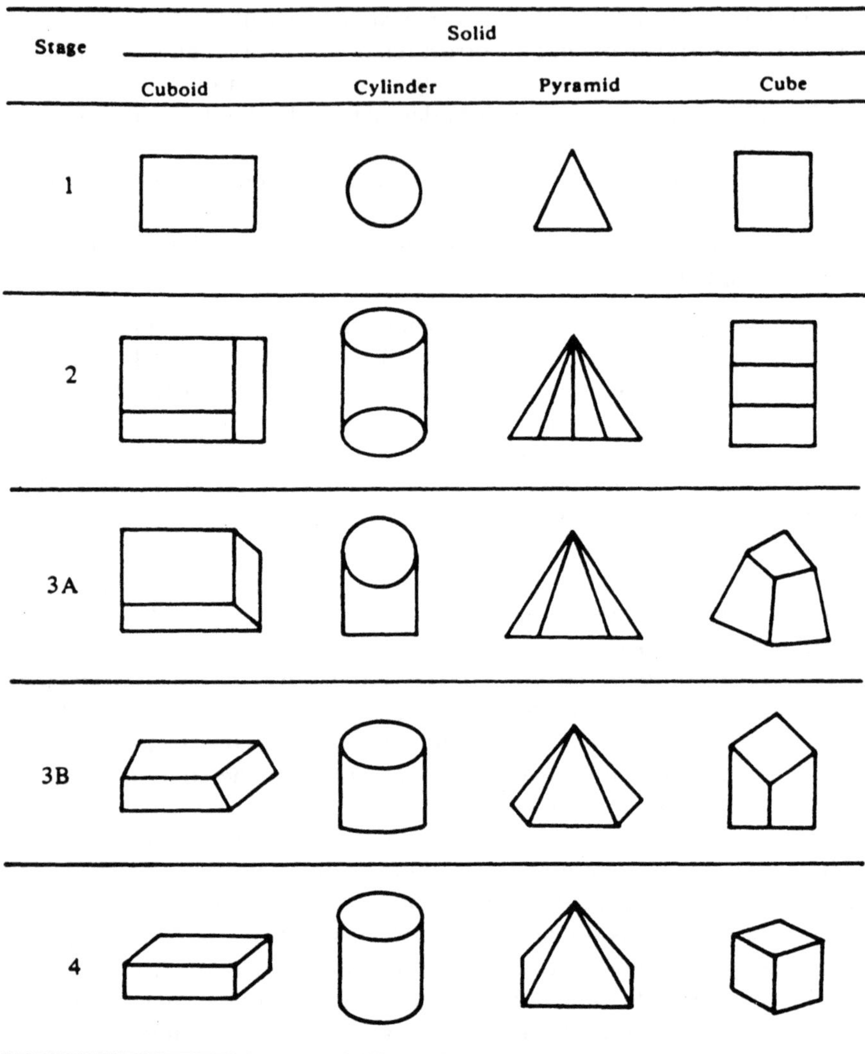

FIGURE 4.3. Mitchelmore's drawings at each stage of solid representation. (Reproduced with permission from Mitchelmore, M.C. [1978]. Developmental stages in children's representation of regular solid figures. *Journal of Genetic Psychology, 133*, 234.)

of unconsciously generated structure, much of it tacit and unrecognized. As he put it,

> the order in which we draw lines and the way we anchor them to one another involves an instantaneous and unconscious judgment about how to minimize errors and avoid too much elaborate control of movement. For example, if you draw a bunch of grapes, you inevitably start with the "front" circle and build the others around it. Otherwise you have to judge the size and position of an arc first, then draw another arc of the correct radius

so that it nicely passes through the end points of the first and so on . . . We tend to think [of such practices] as arbitrary cultural habits, but there is a consistent practical logic behind the cultural variety (p. 64).

In this latter regard, the study of children's drawing and their understanding of pictorial space can be said to be a microcosm of the much larger problems encountered in the study of spatial development.

To a lesser degree, the study of children's left-right identification has continued to attract the attention of psychologists. Both Werner (1926) and Piaget (1928) held that increasing facility with left-right identification is a function of intellectual development. For Werner, children begin by distinguishing their own left and right in terms of body actions. Older children differentiate self from the world around them, and their awareness of left-right relationships becomes a major dimension in their representation of objective space. For Piaget, objective space is achieved when children become able to represent to themselves the left-right relations between objects as they appear from viewpoints other than their own. Younger children's characteristic difficulty with identifying left from right, in other words, is for Piaget a "logical" problem in understanding a relation (Piaget, 1928, p. 109).

In the past twenty years, persistence effort has been expended in the analysis of two-dimensional shape orientation. Among others, Braine (1980), Gibson, Pick and Osser (1962), Thompson (1975), Harris (1975), and Vogel (1980) have noted that left-right relations among objects or within letter-like forms are notoriously more confusing for children than are up-down relations. Fillmore (1971) offered a reason for this confusion in observing that if an object has a left and right, it must also have a front and back and top and bottom. He gave as an example the image of a missile traveling through space. The missile's direction of travel provides it with a front-back dimension but, since there is no reference system to provide it with a top or bottom, it is meaningless to say that it has either a left or right, or that it can make a left turn. Braine (1980) went further when she pointed out that when an object is presented on a horizontal axis, its identity often must be made with reference to some framework provided by the body or the environment:

left and right are inherently relational judgments in the sense that the position of shape must be described with respect to something else . . . The more traditional view, by contrast, assumes that the same cognitive processes underlie left-right and up-down identification, and attributes the greater difficulty of left-right tasks to the less differentiated character of the horizontal axis relative to the vertical axis of space (p. 139).

Children's confusion with left-right identification in three dimensions has also been of interest to clinicians and medical researchers interested in problems of brain dysfunction, delayed laterality, and spatial disorientation. Age-related differences associated with the mastery of left-right identification tasks have prompted some investigators to posit a developmental progression (e.g., Benton, 1959; Corballis & Beale, 1976; Annett, 1985). Recently, Whitehouse, Dayton and Eliot (1980) adapted three identification tasks from Benton's (1959), Harris'

(1957), and Berges and Lezine's (1965) scales, and administered this adapted scale to 578 children between the ages of 3 and 9. The first task required subjects to identify body parts on the left or right sides of their own bodies. The second asked them to make a cross-midline identification of body parts; and the third required identification of body parts on a person facing them. Results from this study were interpreted as providing evidence in support for this developmental progression of tasks.

The Whitehouse et al. study has been described here because it provides a sharp contrast to the descriptive-normative studies undertaken during the first phase. However, although the identification scale may be useful for measuring age-related understanding of left-right relationships, there are still many unanswered questions about how children achieve this understanding, and what effect this increasing ability has upon their representation of events.

The third strand of research, the study of children's understanding of large-scale representation and geographic orientation, has definitely become a research topic in its own right. Between 1980 and 1985 there have been more than 300 publications about children's cognitive mapping, memory for geographic location, way-finding strategies, knowledge of real-world distances in cognitive maps, responses to training in geographical orientation, and similar topics (Eliot, 1985). This outpouring of research interest was prompted, in part, by Piaget's ideas about representation, Ittelson's distinction between object-centered and environment-centered perception, and Pick's studies of children's development of frames of reference. In the past decade or so, this interest has been further stimulated by Siegel's and White's (1975) model, Downs' and Stea's (1977) review, Hart's (1979) naturalistic study, Pick's and Rieser's (1982) further investigations, and Liben's (1981) model. We shall begin this mini-review with a brief description of the Siegel and White model.

The Siegel-White Model

Siegel and White (1975) examined the literature on large-scale representation and, taking a leaf from Lynch's (1960) book, proposed a sequential and hierarchical model. They argued that our everyday existence in a multitude of spaces (home, school, office, etc.) requires us to represent these spaces and their relationship to one another if we are to understand, move about, and interact within and between them. Our understanding of large-scale environments or spaces begins with the discovery and recognition of landmarks in particular places; evolves through the joining of landmarks by paths or routes of travel between landmarks; and eventuates in the coordination of landmarks and routes into survey representations or configurations.

When adults are asked to describe a place, they typically list familiar landmarks. Similarly, young children begin by recognizing recurring landmarks, and gradually come to realize that these same landmarks can be used to organize further knowledge about other landmarks or familiar places. In this sense, landmarks serve both as a strategic focus in our moving about large-scale environments (like a city), and as devices for helping us to maintain our direction of travel. Siegel and White indicate that while young children's ability in recognizing and identifying landmarks evinces comparatively little development, their ability to use familiar landmarks to organize relationships with less familiar ones undergoes significant change.

According to Siegel and White, routes are a form of sequence knowledge that is dependent upon landmark knowledge. Routes may be learned as a sequence of decisions about orientation between landmarks; that is, from knowing that a particular landmark is associated with a specific direction of travel, to knowing that a sequence of landmarks is associated in time with a sequence of orientations. As children mature, their large-scale representations may be conceived as a network of landmarks connected by different routes. Repeated experience with this network results in a gradual shift from knowledge based upon successive associations, to knowledge based upon simultaneous sequences, or routes. With further experience, these networks of landmarks and routes become coordinated into general survey representations of larger and larger segments of the large-scale environment. These survey representations are a form of configural knowledge, which includes multiple landmarks and routes. Presumably, configural knowledge continues to develop if a person travels long distances repeatedly or if there is some reason to use such knowledge.

As indicated, this model intensified interest in children's large-scale representation. Two examples of research prompted by it include a study designed to assess developmental differences in the ability to select and use landmarks for organizing distance information from a walk and, more recently, a study designed to assess the validity of the model's proposed hierarchical sequence of components.

In the first of these, Allen, Kirasic, Siegel, and Herman (1979) examined differences between children and adults with respect to the extent to which they agreed that scenes from a walk had potential landmark value, and the extent to which they agreed that particular scenes depicted critical areas of a walk. All subjects were shown slides depicting what a person would see if walking through a neighborhood on the right-hand side of the street looking forward. The walk itself contained ten actual or potential changes in heading (left or right turns). These changes were deemed to be of critical importance because they entailed decision-making implications for way-finding.

Results from this study indicated that children and adults did not spontaneously select the same landmarks as reference points: adults selected more reference points at actual or potential changes of heading than did children, and older children (fifth-graders) made more accurate distance judgments of the walk than did younger ones (first-graders). These results were interpreted as support for the Siegel and White model.

In the second study, Cousins, Siegel, and Maxwell (1983) assessed the way-finding ability of first-, fourth-, and seventh-grade children around their own school campus. Children were asked to walk three novel and efficient routes; to select photographs of scenes belonging to three routes; to order and to relate metrically those scenes, and to make distance estimates from several sighting locations to six targets within the campus. Guttman scale analysis indicated that 93% of the children exhibited performance patterns congruent with the sequence predicted by the model (i.e., landmark to route to configuration). Despite differences in the accuracy of metric estimations, all children were found to be competent way-finders.

Downs and Stea (1977) argued from a geographic and architectural point of view that environmental learning is the missing link in the chain of connections drawn between spatial problems, cognitive mapping, and spatial behavior. They feel that environmental learning is not something we are taught, but pick up on our own. Moreover, it usually takes place while we are doing something else, operating largely outside the realm of conscious awareness. Our repeated confirmation of existing spatial knowledge is just as important to the environmental learning process as is getting to know new things about a familiar place. In addition to being largely incidental and unconscious, environmental learning experiences are tied in with our plans for doing something in a spatial environment. Our different environments, in other words, form the stage and setting within which our spatial behaviors occur.

These authors claim that environmental learning occurs very early in children's development. They suggest that, as a result of continually decreasing exposure in school to representation of things in geographical space (e.g., maps, aerial photographs, models, etc.) there is a diminution in geographic knowledge as children grow older.

By diminution we do not mean loss of information, but a fundamental change in how this information is known (i.e. stored and processed). Children understand environments less well in a spatial sense because of the reinforcement in school of another specifically verbal mode of information processing (Downs & Stea, 1977, p. 235).

They also suggest that way-finding entails four sequential steps: 1) knowing where you are (orientation), 2) route choice (taking a bearing or direction that will get you where you want to go, 3) route monitoring (maintaining your course by remembering landmarks and their sequence), and 4) recognizing the fact that you have arrived. Although Downs and Stea did not support this sequence empirically themselves, their characterization has prompted efforts by other psychologists to do so.

Hart (1979), for example, undertook a rare and fascinating naturalistic study of children's understanding of place. As he noted, we have learned much about the *how* but little about the *why* of environmental learning. Hart studied different types of ranges of children's spatial activity in their neighborhood as defined by their parents, their place knowledge, their place values and feelings or preferences about places, and their place use. He found that while pre-schoolers could produce fairly detailed maps of the environment around their own homes, they had difficulty with mapping more distant settings. Perhaps more importantly, he underscored the fact that children's learning about what happens in different environments, and about themselves in relation to the environment, is as important as their learning about the environment's spatial properties.

Pick's Model

The study of large-scale spatial understanding continued at an energetic pace. Among those leading this research, Pick and his associates focused upon the relationship between spatially-oriented behavior, frames of reference, and children's representations of large-scale space. In 1981, Pick and Lockman set out these three components in a model.

Whereas Piaget (1956) generalized from the study of object manipulation to the study of spatial layout, Pick and Lockman suggested that the use of such concepts as frames of reference and spatial representation make it possible to generalize from the study of spatial layouts to the study of object manipulation. The key to this reverse development resides in children's spatially-oriented behaviors:

> the spatially-oriented behavior of children ranges from such actions as a young infant's getting its thumb into its mouth, to a preschool child's learning to throw a ball, to an adolescent's hand-sewing of intricate stitches in a homemade garment. The first of these behaviors may be regarded as involving body-body relations; the second of these behaviors involving body-object relations; and the third involving intricate object-object relations.... It would be tempting to propose that there is a developmental trend in performance of spatial tasks involving these three types of spatial relations.... However, even a moment's reflection suggests that the sophistication in all three types of relations increases in the more complex tasks of later childhood and adulthood (Pick & Lockman, 1981, p. 39).

By frames of reference, they refer to the locus or set of loci by which spatial position is defined. *Egocentric* frames of reference refer to spatial positions in relation to loci on the body; *allocentric* frames of reference refer to spatial position in relation to object-object relations. As Acredolo (1977; 1978; 1979) demonstrated, children of different ages are differentially responsive to reference systems defined by the position of objects in a room or by the walls of a room. In her series of studies, Acredolo found an age-related progression in responding to spatial positions from egocentric to allocentric.

Pick and Lockman suggested that spatial representations may be characterized in terms of the operations that a person can perform on spatial information. They point out that spatial representation not only enables us to make inferences about relations between locations, but also to update continually the relation between our own position and all other locations in a space. They emphasize the notion of representation instead of cognitive mapping, because they argue that the focus should be upon the *process* of maintaining spatial orientation rather than upon some metaphoric *product*.

Liben's Model

The construct of spatial representation was further elaborated upon by Liben (1981) in her model of the types and content of spatial representations (see Fig. 4.4). Specifically, she identified three types of spatial representation—spatial products, spatial thought, and spatial storage—and two contents of spatial representation, specific and abstract.

Specifically, three *types* of spatial representation—spatial products, spatial thought, and spatial storage—and two *contents* of spatial representation—specific and abstract—are identified. These are shown below. Although it is difficult to discuss the figure as a whole prior to defining each of the components, several intro-

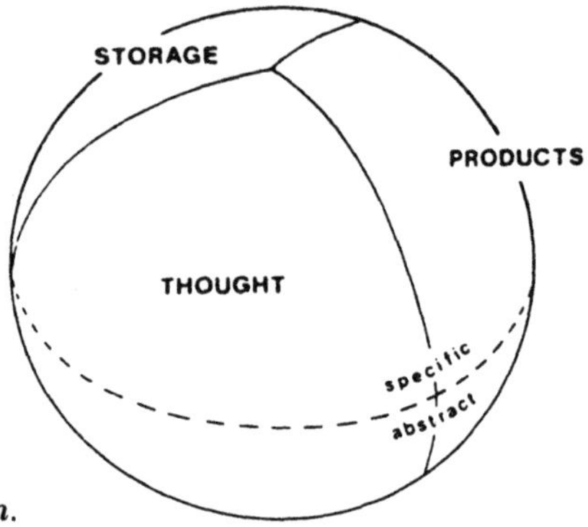

Types and contents of spatial representation.

ductory comments concerning the choice of this particular graphic representation are needed.

First, a sphere is used to accentuate the notion of rotation. Any particular researcher might study the sphere from a different angle. Thus, one investigator might be concerned primarily with spatial thought about abstract spatial concepts, whereas a second investigator might be concerned primarily with spatial storage of information about a specific space (environment). Second, the types and contents are given equal sections to avoid any suggestion that one type or content of spatial representation is more important or more legitimate than another. Third, the representation is unrestrictive with respect to pathways. For example, to go from spatial storage to spatial products it would be possible to go directly, or to pass through (be mediated by) spatial thought. Fourth, the intersections of latitudinal and longitudinal divisions are meant to suggest that each type of representation can be crossed with each content. Finally, boundaries should be thought of as permeable.

FIGURE 4.4. Liben's model and explanation. (Reproduced with permission from Liben, L.S. [1981]. Spatial representation and behavior: Multiple perspectives. In: L.S. Liben, A.H. Patterson & N. Newcombe [Eds.]. *Spatial representation and behavior across the lifespan*. New York, Academic Press, p. 11.)

Liben makes it clear that no one type or content is more important or more legitimate than another. Thus, where one researcher may be primarily concerned with thought about abstract spatial concepts, this work is no more important than the work of a researcher primarily interested in the spatial storage of information about a specific space.

According to Liben, *spatial products* refer to the external products that represent space in some way. These encompass a variety of media, from sketch maps to small-scaled models to verbal descriptions. They refer as much to Freeman's (1980) study of children's drawings as they do to Olson's (1975) or Talmy's (1983) interest in children's linguistic representation of spatial relationships. *Spatial thought*, by contrast, refers to knowledge about spatial relations that we can reflect upon or manipulate in some way. Finally, *spatial storage* refers to any information about space obtained during the lifespan. It differs from spatial thought in that it is typically unconscious, implicit, or tacit. Liben points out that, similarly, language users must possess stored information about their language to enable them to produce grammatical utterances (linguistic competence), but are not necessarily aware of that grammar. She suggests that this analogy applies to the distinction between spatial storage and spatial thought (p. 13).

In addition to describing three types of spatial representation, Liben says that one must consider the *content* of those representations. Where some researchers are interested in determining what knowledge people have about a space in order to move about in different environments, others are more interested in abstract distinctions between knowledge of space and place (Jammer, 1957) or between the perception and conception of space (Cassirer, 1944). She argues that even when a task is developed to assess concrete spatial content, it almost invariably taps abstract content as well. Different methodologies, in other words, may be needed to study the different types and contents of spatial representation. Allen (1985) used Liben's model to review studies of metamemory, linear order representations, frames of reference, and production systems with respect to macrospatial cognition.

During this most recent phase, the study of large-scale representation has been enriched by the sharing of concepts and techniques among psychologists, geographers, sociologists, and urban planners. Although Downs and Siegel (1981) described such studies as being "characterized by a relatively innocent form of low-level empiricism, simple-minded data gathering and classification, co-existing with an abundance of speculative frameworks, each of which is grounded in what amounts to anecdote and common sense" (p. 237), this indictment is perhaps too harsh. Still, the authors insist that better questions must be asked if progress is to be made toward a further understanding of large-scale spatial representation. They suggest "one way to approach these questions is to list a series of methodological 'trade-offs' that might act as organizing themes: performance versus competence, precision versus richness, holistic versus disaggregative, experimental versus naturalistic, vigor versus representativeness in design." They add that "a second approach to asking better questions is to focus upon conceptual issues: the difference between theory and model, the metaphor-analogy distinction, cognitive mapping instead of cognitive maps, large-scale spatial representation versus fundamental spatial concepts, the distinction between modes of representing and mapping, etc." (p. 247). Although not all of these issues have been discussed here, it can be seen that this listing of methodological and conceptual issues represents a considerable advance over

Trowbridge's (1913) and Freeman's (1916) ideas about children's large-scale representation and understanding of cardinal directions!

In the past dozen years or so, there have been only a few models advanced on the scale of Piaget's or Werner's. One example is the highly integrative model set forth by Olson and Bialystok (1983), who state that they have attempted to respond to the

> problem of inner space or spatial cognition, the spatial features, properties, categories, and relations in which we perceive, store, and remember objects, persons, and events, and on the basis of which we construct explicit lexical, geometric, cartographic and artistic representations (p. 2).

Given this concern, their purpose is to describe a form of mental representation that jointly specifies the meaning of objects and events and the spatial information appropriate to their perception. A second purpose is to specify the processes whereby these structural descriptions are "interrogated" to form explicit spatial concepts with their own distinctive meanings. The authors contend that these spatial concepts are then organized at a new level of meaning, which they call *representations of form*, in order to distinguish them from representations of objects or events (p. 3).

Olson's and Bialystok's Model

This model is not easy to understand. Its central idea is that, as in Chomsky's notion of deep and surface structures of language, we construct a deep structured or underlying representation of object, layouts, and events, and then "read off" from these structural descriptions a surface response depending upon the contextual demands. We perceive objects and judge them to be similar to one another if they involve the same structural description. Thus, a structural description

> may be considered a propositional representation of the properties or features and their relations constructed by the mind which permits the recognition of an assignment of meaning to objects. It is similar to a feature list except that the features are not assumed to be merely a list but an ordered, hierarchically organized set of descriptions. Furthermore, structural descriptions are assumed to be constructed from a set of elements that is smaller than the set of objects represented. They, therefore, constitute the language or code for the mental representation of experience (p. 8).

The notion that these underlying structural descriptions have a propositional character is based, in part, upon Palmer's (1975) and upon Pylyshyn's (1973; 1981) ideas. We shall consider Palmer's ideas first.

According to Olson and Bialystok, Palmer (1975) suggests that propositional structural descriptions contain information about features, relations between features, and values on a dimension and encoded at a level appropriate for recognition, discrimination, etc. Using a minimum of parameters, Palmer says that propositions can be constructed to represent any visual display.

> The structural description for a square, in Palmer's system, requires three parameters—location, orientation, and length. This information is given by the more primitive (i.e. component) features of lines and angles of the square which, in turn could be specified. Thus the parts and the relationships between parts of a figure can be retrieved at various levels. The meaning is represented by the higher order relationships, or parameters, which define the shape (Olson & Bialystok, 1983, p. 9).

The advantage of structural descriptions, according to Palmer, is that they can be used to represent various types of information. Further, propositional structural descriptions may encode analogical images upon which analogical operations might be performed. Hence, structural descriptions are not alternatives to images but are an underlying form of representation from which images may be generated (p. 9).

Pylyshyn (1977) has argued that the representation of objects or events is in the form of propositions rather than images. Although he acknowledges that images exist, he says they are non-pictorial in character, just as structured descriptions are non-linguistic. Structured descriptions are constructed by selecting from a "vocabulary of available concepts" to form an appropriate representation. Pylyshyn argues that a computational model using such proportions can account for data from spatial mental rotation problems as well as for theories that account for such data in terms of analogue mental images (Olson & Bialystok, 1983, p. 11).

With this background in mind, Olson and Bialystok argue that, while meaning is important at every point in the construction of structural descriptions, meaning and structural descriptions can be specified separately (independently) but are themselves highly interdependent. Meaning systems reflect the goals, purposes, and intentions of a person; structural descriptions reflect the properties associated with assigning what is perceived to a meaning category. Moreover, where information in structural descriptions (colors, shapes, locations, sizes, etc.) is largely implicit or not attended to as such, the meanings assigned to such properties are explicit. As they explain,

> spatial displays or spatial relations generally are cognized by means of assigning a structural description to that event; to learn about an object or event is to construct a structural description for that event; to recall an event is to retrieve the appropriately stored structural description. Further, each distinctive structural description possesses a distinctive meaning; indeed the latter is an important condition for the construction of the former. Two events which take the same structural description are considered the same, and two events, if it becomes important to differentiate them, that is, if they have different meanings, must be assigned distinctive structural descriptions (p. 15).

Olson and Bialystok further content that children do not perceive space as such, but instead use spatial information in their structural descriptions of objects, layouts, and events. While spatial information is extremely important in the construction of structural descriptions, the perception and explicit cognitive representation of space is derivative. They claim that the progressive explication of spatial information implicit in structural descriptions constitutes what is involved in the development of the perception of space. Language is an important means for making that spatial information explicit (p. 47).

If spatial knowledge is implicit, then how do children come to see the similarity between a balloon on a string and lollipops, given that they have different meanings? Olson and Bialystok say that children do not see circles and lines, but perceive balloons and lollipops: for lollipops, the circle and line are not perceived in their own right but only in the service of recognizing balloons. The distinction becomes one of separating object from form perception, where exactly the same visual information is used but at different levels of meaning. The development of form representation requires that information implicit in object representation becomes explicit and be assigned a meaning, thereby becoming an independent concept. At the level of form representation, balloons and lollipops may be treated as equivalent provided that they are similar at the explicit level of form, as in the class of round objects. Spatial concepts develop when implicit features become detached from their former object meaning and are assigned their own distinctive meanings with their own structural descriptions. At this point, the circle and the line may be perceived in their own right. It is this detachment of form from object that permits explicit spatial thought, and the appearance of new forms, the reassignment of forms, and the representation of this activity in language and drawing (p. 77).

With respect to language, Olson and Bialystok claim that adult representation entails the voluntary assignment of spatial axes and predicates that were implicit and automatic in a child's perception of space. Specifically, spatial conditions

> are constructed on the basis of propositions which relate an argument to a relatum through a spatial predicate, for example
>
> on (book, desk), i.e., The book is on the desk
>
> where "on" is the predicate, "book" the argument, and desk is the relatum. These 'structural propositions underlie object recognition. Some spatial propositions specify the relations between objects such as relative positions. These "relational" propositions' underlie our knowledge of locations. Together, such propositions serve as the mental representation of objects, patterns, locations, and events (p. 234).

This abstraction of ideas from Olson and Bialystok's model is necessarily incomplete for such a complicated model. As the authors point out, their model addresses a very broad range of theoretical problems. Moreover, it provides a basis for interpreting such issues as the relationship between perception and representation, language and visual perception, and imagery and spatial representation. It is unusual in that such problems as pattern perception, spatial imagination, and spatial problem solving are held to involve a common set of mental representations and cognitive operations.

In sum, research on spatial development during this third phase, although still under the influence of Werner and Piaget, became more topically focused as researchers undertook to follow up different strands of interest. During this phase, the study of children's drawing and the study of their large-scale representation emerged as such topical foci, while their difficulties with left-right identification received somewhat less attention. Except for the Olson-Bialystok model, most of the other models advanced to account for spatial development were more topic-specific.

In all three phases summarized here, there have been important advances in our understanding of spatial development. While the first phase was dominated by preoccupations with the collection of normative data about children and comparisons of their behavior with that of adults and animals, the second phase may be characterized by the gradual infusion of experimental methodology and the study of children as children. In the most recent phase, the influence of Werner and Piaget has begun to give way to more topical concerns, and the study of children has become an important component in the examination of development over the entire lifespan. In all three phases, we have made reference to models that possess a spatial dimension, and more specifically to models concerned with spatial development. To conclude, as in the preceding chapter, we will now turn to a consideration of some general and specific issues.

Issues

Despite the emphasis in research on age-related changes in spatial behaviors, most theories have emphasized the emergence of these behaviors and ignored their persistence or decline. As we have seen, the emergence of spatial functioning has been accounted for in a variety of ways. Baldwin described a progression from subjective to objective experience; Gesell accounted for age-related differences in terms of maturational principles; Cassirer described an evolution from action through perceptual to conceptual space; Werner looked at changes in structures from a relative state of globality to states of increasing differentiation, articulation, and hierarchical integration; and Piaget described the achievement of representational space as a shift from perceptual to conceptual cognitive systems, and a progression from reproductive to anticipatory imagery. Bruner accounted for the acquisition of symbolic function through three modes of representation, and Vygotsky described similar age-related shifts in terms of a progression from social to individual planes. More recently, Pick and his associates concentrated upon young children's frames of reference, and Olson and Bialystok accounted for the achievement of spatial cognition in terms structural descriptions. While each of these models has contributed to our understanding of the emergence of spatial functioning in childhood, they tell us little about its decline in old age.

Central to Piaget's account, as we have seen, are the assumptions that his variables develop in a unidirectional, progressive manner, and that the achievement of processes is invariant and irreversible. Presumably, these assumptions hold throughout the lifespan. However, both Papalia (1973) and Rubin, Attewell, Tierney and Tumolo (1973) have reported evidence to support the idea that these processes can reverse in old age, and that spatial representation may be curvilinear across the lifespan. Unfortunately, apart from a few comparisons between young and elderly subjects (e.g., Ohta, 1983; Walsh, Krauss, & Renier,

1981), we know very little about changes in spatial behaviors during middle and old age. As Kirasic (1985) put it in a slightly different context:

At present, this small literature affords only the "non-conclusion" that spatiovisual abilities of elderly adults either remain stable or decline with increasing years. Ironically, even if this issue were resolved, the implications for spatial behavior would remain unknown. (p. 187).

Mandler (1984) made a similar observation when she pointed out:

We do not know either for the child or the adult which kinds of spatial knowledge will increase with experience and which will not. I referred earlier to adults lack of detailed knowledge about many aspects of faces; at some point, further experience does not lead to further detail (p. 452).

Also central to Piaget's account is the role of action in the achievement of representation. If indeed the achievement of cognitive systems affects the representational activity of children, it could be that the activities of the middle-aged and elderly similarly affect their spatial functioning. Although Newcombe, Bandura and Taylor (1983), Lunneborg and Lunneborg (1984), and Olson and Eliot (1986) have studied the relationship between everyday activities and performance on spatial tests, these studies have been faulted for the limited size of samples, the comparatively limited scope of activities investigated, the narrow range of item types in spatial tests, and the failure to examine the simultaneous influence of other environmental variables (e.g., occupational demands, processing styles, self-assessment, physical output, etc.). Nonetheless, these studies constitute the beginning of a promising line of research, which could eventuate in our increased understanding of the effects of occupation and everyday activity upon the spatial functioning of the middle-aged and elderly.

The second general issue, that of sex-related differences, has scarcely been mentioned in this chapter. As indicated in the previous chapter, however, several explanations for the existence of such differences have been proposed in the past decade, including the possibility that they are transmitted by an X-related recessive gene, that they reflect chromosomal or metabolic influences, or that they are related to maturational differences in the brain. Of these various explanations, Waber's (1977; 1979) research on maturational differences in the appearance of sex-related differences has provided some interesting ideas.

According to Waber, the organization of brain functions should be thought of as a growth process that parallels physical growth itself. When conceived as such, the overall maturational rate and the brain's functional lateralization are related such that late-maturing individuals would be more lateralized than early maturers. Since girls mature earlier than boys, adolescent girls perform better on verbal tasks, while adolescent boys perform better on spatial tasks. Sex-related differences, in other words, may be a function of greater right hemispheric specialization in adolescent males, a specialization that leads, in turn, to an advantage in spatial functioning.

Waber's hypothesis, which was supported by her research, was only one example of attempts to account for the emergence of sex-related differences in spatial performance at adolescence (for a more thorough review, see Sheperd-Look, 1982; Hyde, 1981). Unfortunately, many of the more recent studies have involved children who are at least fourteen years of age. In addition, the sex-related differences that have been documented can account for only a small percentage of population variance. With respect to this last point, Hyde (1981), for example, has calculated that the contribution of sex-related differences to performance on spatial tasks is somewhere between 1% and 5% in explaining the total population variance on spatial tests.

Maccoby and Jacklyn (1974) reviewed an extensive research literature and concluded that sex-related differences in performance on spatial tasks appear from adolescence onward. However, despite this, and despite the explanations advanced by Waber and others to account for these differences, there is still considerable controversy over when such differences appear. On the one hand, Linn and Petersen (1985) found no significant sex-related differences in their study. On the other hand, Johnson and Meade (1987) found evidence for differences appearing as early as the primary school years and increasing in magnitude through age eighteen. Clearly, a great deal more research is needed before we can hope to resolve such questions as the existence, emergence, and extent of sex-related differences in spatial functioning.

The third general issue, the trainability of spatial behaviors, has been referred to only indirectly in this chapter. As Beilin (1971) and others have noted, much of the early Piagetian research in the United States was undertaken to replicate and accelerate the achievment of cognitive systems as presented in Piaget's developmental account. The results have been inconsistent and inconclusive. For example, with respect to training children to respond to Piaget's three-mountain task, Eliot (1966) was unsuccessful while Cox (1979) did manage to train children to imagine a viewpoint other than their own. When Hughes and Donaldson (1979), in a variation of the task, presented two policemen dolls and asked children to hide a third doll from them both, the children had no difficulty, although they did have difficulty with the original Piagetian task.

Lean (1981) compiled an extensive listing of studies that looked at the training of spatial abilities. He concluded that:

> The evidence . . . indicates that these various spatial skills are trainable given the appropriate experiences. Brief training with pictorial materials is sufficient to induce pictorial depth perception; relatively brief practice is sufficient to improve subjects' performance on spatial test items; the teaching of various spatial conventions and exercises with diagrams helps to improve geometry performance; and sufficiently long experience appears to improve performance on spatial tests similar to those done in the drawing course . . . The more successful training studies were carried out with younger children (5-year-olds, 5th and 7th grades), whereas studies with older subjects have had less success (pp. 10–16).

Lean's conclusions notwithstanding, it remains unclear from this literature the extent to which different kinds of training are generalizable to other tasks, to which the training results in any permanent effects, to which training may contribute to understanding of mathematical or scientific concepts, or to which training improved spatial ability or the ability to take spatial tests.

From the viewpoint of education, Bishop (1980) raised some interesting questions about the use of manipulative materials in the training of spatial skills. He quotes the following passage from Sherman with respect to teaching spatial skills:

> Research needs to be directed towards factors affecting the development of spatial skills not only in the early years, but even in the adult years. We do not assume that an illiterate adult should be written off as unteachable, nor should we assume that adults cannot improve their spatial skill. Methods of teaching this . . . need to be devised, and their feasibility and advisabilty evaluated . . . For a swimmer with a weak kick we provide a kickboard and opportunities to develop the legs. We do not further exercise the arms (Sherman, 1979, pp. 26–27).

Bishop then goes on to ask how strong does an ability need to be before it can be regarded as a strength? Conversely, how weak does it need to be before it is considered a weakness? Specifically, he suggests that a teacher's or trainer's level of spatial ability ought to be taken into account as part of the experimental teaching method employed.

The role of instruction in affecting spatial abilities has been alluded to in this chapter with respect to Olson's notion of preformatory acts, and Vygotski's "zone of proximal development." Recall that Olson argued that direct instruction could provide children with guidance about choices of action in the performance of different tasks in different media, and Vygotski argued that his "zone of proximal development" was consistent with the belief that what children can do with the assistant of others may be more indicative of their mental development than what they can do alone. Certainly, both arguments are noteworthy with respect to efforts at training spatial ability at all ages.

The fourth general issue, that of the role of imagery in spatial thinking, has already been mentioned in this chapter. However, imagery has been ascribed different roles by different theorists. These range from Baldwin's conception of imaginative schemata as a means of uniting logic with thought, to Piaget's classification of different kinds of imagery and their development as a cognitive subsystem, to Bruner's employment of imagery as a transitional mode between enactive and symbolic modes of representation, to Olson's and Bialystok's conception of structural descriptions as forms of representation from which images may be generated. From this range of interpretations, it is often difficult to know whether to regard imagery as an ability, a mediating process, a mode of representation, a skill, or a stimulus attribute.

An ongoing developmental controversy is whether imagery impedes or facilitates thought in childhood. As Galton (1880), Bruner (1966) and others have noted, children appear to use imagery more than adults, and eidetic imagery, common in childhood, appears to fade in adolescence about the same time that spatial ability emerges (Haber & Haber, 1964). For some researchers, high imagery in children is a valuable component of thought (Anderson and Bower, 1973; Kosslyn, 1980); for others it is regarded as a primitive way of representing information that impedes more symbolic or abstract thought (Kuhlman, 1960; Brooks, 1967; Stafford, 1972). Brooks (1968), in particular, reported evidence that imagining and perceiving do conflict with one another, at least in some conditions.

Hollenberg (1970) conducted a very interesting study using tests of spatial ability as measures of imagery. She predicted that children with high imagery would learn labels of objects faster than those with low imagery, while the reverse pattern of results was predicted with respect to the acquisition of concepts. In an extreme-groups design, 64 children from grades 1 to 4 were asked first to attach four nonsense syllables to pictures. Next, they were given four sets of pictures in which each picture represented other examples of the concepts underlying the first four pictures. The children were then asked to anticipate the "names" of the new pictures.

Her results showed that high imagery subjects did indeed learn labels faster, especially in the first grade. Her second hypothesis, that high imagery children are slower in acquiring concepts, was also supported. Hollenberg interpreted her results as evidence that imagery interferes with learning concepts, because to grasp a concept it is necessary to look under the visual appearance of something for its critical attributes. High imagery children were slower to learn concepts because they tended to fixate on the "surface structure" of appearances. It is interesting that Twyman and Bishop (1972) and later Lohman (1979) pointed out that this conclusion did not take into account the gradual increase in children's ability to reason about figural problems—that in fact a good reasoner with poor imagery might outperform poor reasoners with good imagery. In any event, the relationship between perception, imagery, and reasoning during childhood and adolescence has become more complicated in the last twenty years.

It is unfortunate that so many studies of imagery have used spatial tests as measures, especially as Di Vesta, Ingersoll and Sunshine (1971), Schekels and Eliot (1983) and others have found that imagery and spatial ability load on separate factors when imagery is measured by self-reporting questionnaires. Not only is there a clear need for the development of new imagery measures, but there is also a need to measure different kinds of imagery across the lifespan.

The fifth general issue, the types of processing required by spatial tasks, has been the central concern of several theorists mentioned in this chapter. Early notions about processing regarded it in terms of the sequential solution response to a given problem situation, as in Olson's pattern of decisions affecting children's drawing. More recent approaches have incorporated information-processing connotations, with processing occurring in a programmed or given sequence of response decisions or actions to a fluid problem situation. In the latter sense, the concept of processing carries with it the implication of a comparatively limited number of sequences for an unlimited variety of problems, with types of processing taking on many of the characteristics formerly attributed to traits or abilities. Whereas earlier notions about processing emphasized organizational characteristics of responses, later notions emphasize the task demands of different stimulus situations.

This change in the use of the concept of processing is evident in our review of develomental models. On the one hand, Baldwin's construct of processing via cortical coordination, Gesell's idea of complementary processing, and

Werner's progression of sensorimotor-perceptual-conceptual stages, each reflecting an increased degree of part-whole articulation and means-ends differentiation, can be said to emphasize the organizational basis of responses. Piaget's invariant processes of assimilation and accommodation may be said to mark a halfway point in the change of the concept of processing, emphasizing as it does the interaction of both stimulus and response. More recent interest in the directive function of language on spatial thought, and in way-finding and route-monitoring through large-scale environments, emphasizes influences that affect the ordering of decision-making activity in the solution of tasks, whether they be sensory, perceptual, inferential, or representational in nature. As will be seen in the next chapter, several process models have been proposed which, for the most part, have yet to be applied to the performance of spatial tasks or activities across the lifespan.

The last of these general issues, the effect of stimulus dimensionality upon spatial performance, has been directly addressed by Ittelson (1973) and by all those who have subsequently investigated differences in responding to figural, object, and large-scale spatial tasks. The literature that sets out the characteristics of these three levels of stimulus dimensionality is very large; the literature on the effects upon performance of similar tasks from a developmental point of view is small but rapidly growing. Among several excellent reviews are those by Weatherford (1981), Hazen (1983), and Mandler (1984).

A related issue, mentioned earlier in the chapter on philosophical conceptions of space, is the extent to which our spatial representations of different surroundings correspond to the actual layout of geographical features and locations. As Downs and Stea (1977) noted, our environmental learning is not something we are taught, but pick up on our own. Since our representations of large-scale environments are affected by our cognitive status, the nature of our experiences, and our feelings about different places, Cohen (1982) says that our cognitive maps of such environments are not likely to be isomorphic with physical space. Presumably, the same reasoning may apply to our representations of object arrangements and even two-dimensional configurations. What remains to be discovered is whether different patterns of correspondence exist under different conditions for different ages across the lifespan.

With respect to issues that are more specific to the developmental approach, several have already been indicated indirectly in our description of various models. A persistent issue has been the problem of accounting for apparent stability and changes of behavior, both within and between age groups. Although some significant advances have been made, there is still considerable debate over what constitutes a "stage" in descriptions of spatial development, and how perception, imagery, and representation interact within stages to maintain or further that development. As Pellegrino, Alderton and Shute (1984) observe,

> Data from longitudinal studies typically reflect stability in individuals' performance relative to each other while disregarding any absolute changes such as all individuals being able to solve more problems or more difficult problems. Data from test-retest reliability studies typically reflect stability in both relative and absolute performance but without

> any intervening experience that might be expected to produce a general or differential change in absolute performance (p. 30).

In other words, in some instances the methodology and the measures we use to assess developmental differences are still too contrived and crude to enable us to tease out subtle expressions of stability and change in the comparisons of spatial performance by different age groups.

Two brief examples should help to clarify this last observation. First, in keeping with Gesell's and Werner's preoccupation with the efforts of individuals to self-regulate their behavior and, to some extent, their own development, we need to ask how we may encourage subjects to exhibit self-regulation patterns in testing situations where, by virtue of the testing situation, they cannot be allowed to self-regulate.

The second example is even more specific. As Smothergill (1973) noted, Piaget defined topological space in terms of proximity, which precedes projective and Euclidean space. However, as Mandler (1984) pointed out

> it is difficult to separate topological from Euclidean aspects of spatial representations . . . a difficulty which is compounded by the persistent finding that improvement on both topological and Euclidean measures occurs throughout childhood. [Since] much of the superstructure of spatial knowledge at all ages is apt to be topological in character, organized around landmarks, linear orderings, and relations of enclosure and relative proximity, [we should] value topological encoding rather than devalue it (p. 452).

What are needed, in other words, are testable models of what it is like to live in a topological spatial world.

Another persistent issue concerns the agency of developmental change. Perhaps, as Liben (1985) has suggested, we have overworked the role of action in knowing, although Cassirer, Piaget, and others found it a useful (and necessary) counterbalance to the passive organism of behavioral conception. Surely action is a critical component, but so too may be the directive functions of language and instruction, as Bruner as well as Olson and Bialystok have indicated. As methodology in developmental studies becomes more sensitive, perhaps formulae will emerge that will provide some way to weigh different combinations of feeling, action, and the influences of language, context, and instruction upon the spatial performance of subjects at different ages in different assessment situations.

In this chapter, the developmental approach to the study of psychological space has been summarized in terms of three phases of research interest. Within each phase, several models of development have been described, some of which include a spatial dimension, others of which are more narrowly focused upon spatial cognition. In addition, three research strands were described across the three phases: namely, research in children's drawing, left-right identification, and development of large-scale spatial representations. Following this summary, six general issues were discussed, as well as two specific developmental issues. In the next chapter, a similar range of models and issues will be presented from the experimental approach.

5
The Experimental Approach

The study of spatial perception, imagery, and spatial information processing is primarily the concern of experimental and cognitive psychologists who assume, for the most part, that our knowledge of the physical world comes through our experiences, and our experiences come only through our senses. Typically, the experimental and cognitive researcher studies individual differences by observing, manipulating, and quantifying behavior in terms of speed and/or accuracy of responses to given stimuli presented under controlled conditions. Individual differences, however, are not the primary concern since major efforts are directed toward inter-stimulus differences where response patterns provide causal rather than correlational explanations. When individual differences are studied in perception, they are usually interpreted in terms of sensory systems or response thresholds, in imagery they are usually interpreted in relation to other knowledge systems, and in spatial processing they are most commonly interpreted in terms of computational models of information processing.

It must be emphasized that the experimental or cognitive study of spatial perception, imagery, and spatial processing constitutes only a very small part of the behavioral literature and, further, is only one of several domains of interest to experimental and cognitive theorists. Those interested in spatial topics tend to ask such questions as how we acquire or learn about an apparently stable world of tangible objects from out of the continuous flux of changing stimuli, or how different kinds of imagery affect our perceptions and thinking about the external world, or how different preferences for processing and strategy choices are influenced by different stimulus demands. The experimental or cognitive theorist identifies problems or tasks as interesting if they can be tested by hypotheses generated from theories or models; if they can be presented in observable and replicable conditions; or if they can be structured to yield quantifiable responses or behavioral patterns that can be simulated on a computer.

As indicated in Chapter 2, the ideas that were historically important to the emergence of the experimental approach can be traced from Aristotle through the British empiricists and, more recently, to Darwin's theory of evolution and to laboratory research with animals. As it has often been pointed out, the term "experimental" approach refers more to an attitude about *how* behavior should be studied, rather than to *what* behaviors are most worth investigating. The emphasis is clearly upon overt and observable behaviors rather than upon rational or mentalistic constructs.

With respect to spatial interests, it was Berkeley (1709) who not only articulated the empiricist viewpoint about how we come to know the physical world, but also contributed to our thinking about spatial perception. Recall that he argued that we obtained our knowledge of the physical world from direct but unrelated sensory impressions, from images or faint copies of previous impressions in memory, and from the transformation, decomposition, and recombination of impressions and memories in associative thought. Berkeley's three-part description is useful not only because it neatly summarizes his empiricist viewpoint, but also because it suggests a way to organize our survey of models and research. That is, rather than presenting the experimental approach here as a single brief history, Berkeley's description suggests that it should be organized in terms of three parallel histories: namely, the study of spatial perception, of imagery, and of spatial processing from the information-processing perspective.

As indicated, Berkeley put forth the argument that visual sensations, by themselves, are not sufficient to account for spatial perception. Since the retinal image of the eye is only two-dimensional, he pointed out that it is impossible to perceive directly the size, shape, mass, or location of physical objects in a three-dimensional world. Rather, meaningful perception occurs when sensations as signs or cues are combined with images or faint copies of previous sensations in associative thought through repeated experiences.

From Berkeley onward, spatially oriented behavior was studied in terms of the so-called spatial senses: vision, audition, and touch. Empiricists in particular undertook to look for cues in each sense separately or to study these cues separately. As Freedman (1968) put it,

> in vision we have monocular and binocular cues for depth perception; relative size, interposition, linear perspective, aerial perspective, movement parallax, accommodation, convergence, stereoscoptic vision; and we have a great many experiments focusing upon one or another of these cues . . . The attempt has been generally to hold all else constant, manipulate input information relevant to one type of cue, and measure performance. Experiments are performed in small, dark quiet rooms with the observer seated motionlessly at the apparatus; he may even be restrained to be sure that "movement" does not contaminate the results of the sensory study (p. 1).

The early sensory psychologists and physiological psychologists regarded the physical world as a sensory continuum, and analyzed experience in terms of sensations and feelings—these being the simplest possible contents of mind that, when brought together by association, constituted consciousness. As Miller and Johnson-Laird (1976) described it,

> much attention was given to isolating and measuring the different attributes of sensation, and competing theories were advanced about the neurophysiological coordination of sensory experience (p. 14).

By the early 19th century, the more widely accepted view was that sensations provide the raw material of experience, and that our experiences of objects and space depend upon our combining, ordering, and uniting sensations into perceptions. With the passage of time, those interested in spatial perception came to hold the view that our perception is never completely determined by external stimuli; that it goes beyond stimuli and is abstracted from sensation. By the turn of this century, it was widely believed that perception is more than the coordination of sensory impressions, and that "we use the various sensory excitations as cues to bring a perception onto the stage of consciousness" (Boring, 1946, p. 99). In the first section of this chapter, the sensory-perceptual component of Berkeley's tri-part description will be elaborated upon further, and descriptions of models presented.

The second component, imagery, has also had a distinctive history. The role of imagery in thinking was first articulated by Aristotle, who wrote

> We cannot think without imagery, for the same thing occurs in thinking as is found in the construction of geometrical figures. There, though we do not for the purpose of proof make any use of the fact that, for example, a triangle has definite fixed limits, we nevertheless draw it of a determinate size. Similarly in thinking, though we do not think of the size, yet we present the object visually to ourselves as a quantum (i.e. of a certain size), though we do not think of its size or quantum (De Memoria, 11 4496).

Although there was considerable speculation about the role of images and thought through the centuries after Aristotle, his general view was widely accepted by many philosophers, and the notion of imagery as a function of memory was embraced by many early psychologists.

In keeping with Berkeley's account of spatial perception, psychologists of the empiricist persuasion maintained that our observations of three-dimensional space are not entirely visual, but instead are kinesthetic images, or memories of previous movements and sensations that gradually become associated with visual stimuli through repeated experiences.

According to Piaget and Inhelder (1971), psychologists in the early years of experimental psychology regarded the image in one of three ways:

> 1) it was viewed as a direct product not only of thought but of sensation, and it was held to be a residual trace of the latter; 2) it was regarded as one of the two fundamental elements, the other being association. Thought itself was conceived of as a system of association of images; and 3) it was thought to be an accurate copy of objects or events, and not a symbol subordinate to some operational activity (p. xiv).

Of these three interpretations, perhaps the most popular was the third: that knowledge is a copy of reality in which the image itself is the idea.

At the turn of the century, the first systematic study of the relation of imagery to thought was undertaken by psychologists of the Wurzburg school. Oswaldo Külpe, a German psychologist, discovered what became known as "imageless thought." Kosslyn (1983) described one of Külpe's experiments as follows:

> First lift a book, put it down, then lift a glass. Which object is heavier? What was surprising about Külpe's experiment was that although the subjects could answer the question easily enough, they had no idea how they made their judgments. They were aware of

plenty of images and sensations, but these did not seem to be the basis for their answers—the judgments just seemed to pop into mind full-blown, unguided by conscious processing. Külpe concluded from this that not all thought is accompanied by mental images and thus introspection alone will not reveal everything we want to know about the mind (p. 38).

As a consequence of similar experiments, and the fact that imagery has no physical correlate in behavior, its study was generally neglected during the great behavioral eclipse from 1920 to the 1950s. Only a handful of researchers (Averling, 1927; Bartlett, 1932; Bowie, 1935; Pear, 1927), most of whom were British, continued to study imagery and its relationship to thought. The more recent resurgence of interest on the subject of imagery is the subject of the second section of this chapter.

The third component, association, also has deep roots in Greek philosophy. As Woodworth and Schlosberg (1954) noted,

Aristotle in particular attempted to account for the train of thought by arguing that one idea would be followed by another which was similar or contrasting, or which would have been present together with it in one's past experience (p. 43).

It is interesting that in the many books tracing the history of association, Aristotle is frequently mentioned in connection with an emphasis upon sensation, imagery, and the three attributes of association—contiguity, similarity, and contrast, although in the later works by the British empiricists, as Baumrin (1975) points out, there is little such attribution. Of the British empiricists, perhaps Mill (1829) described the position on association most succinctly when he wrote

Thought follows thought; idea follows idea, incessantly. If our senses are awake, we are continually receiving sensations, but not sensations alone. After sensations, ideas are perpetually excited of sensations formerly received; after those ideas, other ideas; and during our whole lives, a series of those two states of consciousness, called sensations and ideas, is constantly going on. I see a horse; that is a sensation. Immediately I think of his master; that is an idea. The idea of his master makes me think of his office; he is a minister of state; that is another idea . . . and I am led into a train of political ideas . . . (p. 462).

Although the founding of experimental psychology is commonly attributed to Wilhelm Wundt, who established a laboratory in Germany in 1879, it was Ebbinghaus's (1885) experiments that established beyond a doubt that an experimental analysis could elucidate thinking, and set the stage for subsequent efforts in the United States to search for laws whereby associations are formed and remembered. Until the 1960s, there was widespread faith that fundamental laws of thinking could be found through associationistic analyses of performance in simple procedures such as the paired-associate learning procedure. Empiricist outlooks and methods prevailed, and the idea of association contributed significantly to a fascination with learning theory at the expense of both perception and imagery.

During the 1960s, as chronicled elsewhere, psychologists became less concerned with the determinants of association formation and more interested in inner capacities and processing mechanisms. Fewer studies were conducted in which the subject's performance was conceptualized as a matter of passive or automatic stimulus generalization based upon the number and saliency of sensory elements shared between stimuli. Instead, more and more studies were undertaken in which performance was based upon the specification of postulated cognitive processes, and in terms of process models that used the computer as a metaphor for mind. A number of these information processing models, particularly those involving spatial processing, will be described in the third section of this chapter.

Before turning to these three components, it must be emphasized again that publications about spatial perception, imagery, and spatial processing, when taken together, still constitute only a comparatively small portion of the accumulated literature of experimental psychology. Indeed, Berkeley notwithstanding, Hochberg (1957) wrote that

> the perception of space, depth and distance is frequently treated in textbooks as a solved problem. Despite the fact that some restricted areas of precise and applicable knowledge exist, however, the basic problems in this area are completely unresolved (p. 83).

Now, thirty years after Hochberg's observation, at least some of these basic problems are being addressed and new problems encountered. The three components discussed in this chapter will make reference to "traditional" and Gibsonian ideas in the section on sensory-perceptual space; to Paivio's, Shepard's, and Kosslyn's models in the section on imagery; and to Egan's, Carpenter's and Just's models in the section on spatial processing.

Sensory-Perceptual Space

As some background to the study of sensory-perceptual space has already been given, we need only note in passing a few significant events that marked the emergence of psychophysics from the philosophy of the empiricists. Bell (1811) and Magendie (1822) separately reported finding that the ventral and dorsal roots in the spinal cord serve motor and sensory functions, respectively. In 1850, Helmholtz demonstrated that it was possible to measure the velocity of the transmission of nerve impulses. And in 1860, Fechner published a book in which he described three experimental methods for determining sensory thresholds. By the latter half of the 19th century, the study of sensation and sensory systems was attracting an increasing number of investigators, partly because works like Fechner's appeared to resemble physiology and its brother, experimental physics. As Hochberg (1962) observed, it was the putting together of sensory units which made Fechner's work look like science.

During the 1890s, attempts were made to treat visual shape as an independent attribute of sensation, but such efforts to reduce perception to elementary sensations were not very successful. As Ittelson (1960) noted,

> the early perceptual work fell within what has come to be known as the tradition of stimulus determinism: the view that the external stimulus determines the perception. Emphasis

was placed on a detailed analysis of the characteristics of the external stimulus, and the search was for universal laws relating the physical environment and the subjective experience (p. 6).

By the turn of the century, theorists interested in perception were also thinking in terms of the perceiver using clues to form inferences about the nature of external objects, thanks to Helmholtz (1925). In the first half of the present century, a variety of theories of perception were advanced to account for different aspects. Many of these theories had little to say about visual spatial perception per se. As Miller and Johnson-Laird observed, Allport (1955) organized this variety of perceptual theories in terms of six major groups:

the classical theories of sensation and associations; the configural approach of gestalt theory; the topological theory of life spaces with their vectors, valances, tensions, and equilibriums; a neuropsychological approach through cell assemblies and phase sequences; motor theories involving sensory-tonic systems and set; adaptation-level theories emphasizing perceptual norms and frames of reference; theories based upon probabilistic weighting of cues or on transactions with the environment; motivational theories of directive states and perceptual defenses; theories that perception is determined by the relative strengths of the perceiver's hypotheses about the world; stimulus-response theories of the role of learning in perception; cybernetic theories involving coding, negative feedback, and information-processing; Allport's own outline of a general theory of event structure. This catalog is a reminder both of the complexity of the problem and of the variety of approaches to it that one might take (Miller & Johnson-Laird, 1976, p. 33).

My purpose in commenting upon the evolution of sensory-perceptual theories and in listing the different kinds of theory extant in 1955 is not to provide yet another history of experimental psychology or perception, but to sketch a background for a discussion of three models (or better, interpretations) of spatial perception in the experimental tradition. As will be seen, each of these models differs in terms of the phenomena included in the construct of perception, and each draws upon a somewhat different heritage within the experimental approach. The first two, Carr's (1935) and Howard's and Templeton's (1966) are representative of the "traditional" approaches to spatial perception; the third, Gibson's (1950; 1966) has been described as a radical reformulation of psychophysics.

Carr's Model

Carr (1935) introduced his book with the statement that the term "space perception" is unfortunate because it suggests that space is something we perceive. Instead, in line with a view repeatedly mentioned in the second chapter, he argued that

only objects are perceived, and these objects possess a number of attributes: qualitative, intensive, and spatial. The term refers to the perception of the spatial attributes of objects, viz., their size, shape, stability, motility, and their distance and directional locations in reference to each other and to the perceiving subject. Space as distinct from these spatial attributes is a conceptual construct (p. 1).

He goes on to define the problem of spatial perception in terms of the spatial cooperation of the senses. He noted, in particular, that

> the various senses co-operate quite successfully in giving us knowledge of the spatial characteristics of objects, such as their location, their size and shape, stability and motility. As a rule, we see objects where we hear them and hear them where we see them . . . The co-operation of the senses with respect to the spatial attributes of objects is but part of the wider question of the co-operation of the senses with respect to all other sensory attributes. According to empirical theory, a child will naturally regard the auditory and visual experiences of any object as two spatially unrelated objects, rather than as two attributes of one object with a common location . . . Are these senses totally unrelated at first, and is the child forced to learn this relationship in the course of experience (p. 8)?

After reviewing the results from experiments by Stratton (1896; 1899) and others, Carr argued that

> the successful co-operation of the senses is thus an incidental by-product of the three systems of association (audition, somaesthesis, and vision). The three senses co-operate successfully in perceiving the location of an object whenever the three stimuli elicit a common localizing response that is adapted to the manipulation of the respective sensory aspects of that object (p. 57).

He goes on to explain that

> we neither hear nor see the distance between an auditory object and ourself. We merely hear the object and interpret its spatial relation to ourself as a visible object. The spatial co-operation of vision, hearing, and somaesthesis is due to their direct connection with a common system of localizing responses (p. 93).

In the remainder of his book, Carr describes experiments in auditory space perception, stereoscopic visual perception, studies of distance and direction perception, and experiments of motion perception, many of which were conducted at the turn of the century, and all of which support his interpretation of spatial perception.

Howard and Templeton's Model

Howard and Templeton (1966), writing thirty years later, can be said to have extended Carr's interpretation with the description of a large number of studies of sensory and perceptual awareness of body and environmental features. Among the topics they reviewed are experiments on the judgment of eye level, the effects of instructions and set upon perception of the apparent vertical, accuracy of movement in different directions, prismatic displacement, and so forth.

According to Howard and Templeton, spatial behavior is constrained by body limitations and by the nature of the physical world. In everyday environments, body movements, their sensory consequences, and external stimuli are patterned because of the way the body and the world are constructed. Spatially coordinated behavior is only possible because there are many predictable features of the body's structure and the structure of the environment. The typical environment, they argue, consists of solid objects in more or less consistent relationship to one another, with some objects that change their position. Adequate spatial behavior implies a matching of one's behavioral repertoire to the given constraints of one's body and the world.

The most fundamental body constraints are provided by the mechanical structure of the neuromuscular system. Howard and Templeton describe the system in terms of four categories of constraint.

> Each joint has a limited direction and a range of movement, and every movement is controlled by the contraction of particular muscles. A second category includes sensorimotor constraints (postural reflexes). A third category is motor-sensory. A fourth category includes intra- and inter-sensory subcategories of environmental constraints. The so-called "perceptual constancies" are examples of ecological invariances (invariant relationships between retinal-image size and object distance). When considered within a single modality, they are intrasensory constraints; when combined they serve as intersensory constraints. Whereas the learning paradigm for the behavioral mapping of self-produced movements is that of operant conditioning; the paradigm for intermodal spatial mapping is classical conditioning (p. 10).

Both Carr's and Howard's and Templeton's interpretations can be described as "traditional" because both are based upon the assumption that sensory elements are welded together to form our spatial perceptions. Also, both are based on the assumption that our sensations are enriched or added to through the coordination of senses, memory images, and the associative process. Both account for change in terms of some kind of addition or accrual to the sensory input from points of stimulation.

Gibson's Model

In 1950, Gibson proposed an alternative formulation based upon perceptual psychophysics. As he put it:

> At the beginning of World War II, the theoretical problem of space perception became a practical problem almost overnight. The skills of aviation began to be of vital interest to millions of individuals. The abstract question of how one can see a third dimension based only on a pair of retinal images extended in two dimensions became very concrete and important to the man who was required to get about in the third dimension . . . The theory of the binocular and the monocular cues for depth, perfected eighty years before by Helmholtz, could explain how a pilot might see one point as nearer than another point. But the pilot was not looking at points of color in a visual field; he was typically looking at the ground, the horizon, the landing field, the direction of his glide, not to mention several instruments, and visualizing a space of air and terrain in which he himself was moving—very fast and possibly in a cold sweat (p. 59).

At the beginning of World War II, there was an urgent need to understand the perception of depth and distance from the perspective of a pilot. The critical task of estimating height from the ground when landing, for example, became particularly important. Many studies were undertaken based upon what psychologists had learned from "traditional" studies, but unfortunately these experiments failed to clarify the problems faced by pilots when actually flying or landing airplanes. As one of the psychologists who was responsible for much of this research, Gibson set out to determine why this research was inadequate and, by the end of the war, he came forth with a wholly new approach to the study of spatial perception:

> The basic idea is that visual space should be conceived not as an object or an array of objects in air but as a continuous surface or an array of adjoining surfaces. The spatial character of the visual world is given not by the objects in it but by the background of the objects. It is exemplified by the fact that the airplane pilot's space, paradoxial as it may seem, is determined by the ground and the horizon, not by the air through which he flies. This conception leads to a radical reformation of the stimuli or cues for depth and distance. Instead of investigating the differences in stimulation between two objects, the experimenter is led to investigate the variations in the stimulation corresponding to a continuous background (p. 7).

Unlike Carr or Howard and Templeton, Gibson argued that it is possible to perceive space with vision alone, that to perceive is to differentiate (to see a difference), and that visual perception becomes richer through differential responses, not through an elaborate central organizing process. He insisted that all relevant information is present in visual stimulation, and that we perceive things spatially in terms of ordinal arrangements, gradients, density and textural changes, frontal and longitudinal surfaces, and invariants under transformations—all features of the stimulus pattern itself and represented in the retinal image as higher-order variables of the stimulation (p. 147).

Like Piaget, Gibson elaborated and revised his ideas over more than thirty years. Although these ideas are fascinating (and controversial) in their own right, for the purposes of this book we shall concentrate instead upon the distinctions and propositions that Gibson put forth in 1950.

Gibson began his analysis by distinguishing between the *visual field of sensation* and the *visual world of perception*. According to him, the visual field differs in a number of important ways from the visual world. First, the visual field is a scene from a given perspective, oriented in terms of visible margins of limits or boundaries. The visual world, by contrast, contains familiar, ordinary objects, perceived against a continuous, unbounded background. Second, whereas the visual field is confined to approximately 180 degrees of regard and is unstable in that shapes or objects are deformed whenever the observer moves, the visual world surrounds us for the full 360 degrees, and is stable in that neither the objects nor the spaces between them appear to change their dimensions as one moves. Third, the visual world depends upon the coordination of impressions which, by themselves, are devoid of meaning, but the visual world is filled with objects that have meaning. Gibson (1950) argued that too much attention had been paid to the study of the visual field of sensations, and not enough to the visual world of perceptions (p. 166).

Gibson undertook to reformulate the study of the visual world. As he observed, studies got off to a poor start because theorists in the 18th and 19th centuries conceived of the issue of how we perceive the physical world within notions of a Newtonian universe. The Newtonian universe, as we have seen, is an abstract, empty Euclidean space made up of points, lines, and intersecting planes, and defined by the three dimensions of the Cartesian coordinates. By contrast, Gibson held that our perception of the visual world is of a space filled with familiar objects in order to be visible, not empty air. As he put it:

> the visual superposition or overlapping of surfaces is an important type of depth perception, not a cue for depth perception. The difference is steps, not cues; steps in the density of textures, steps in the deformation of textures, and steps in the binocular disparity of textures. These steps are proportional to the physical difference in depth between the two surfaces in question (p. 228).

Gibson himself summarized the main principles of his model. He began by arguing that a fundamental condition for seeing the visual world is as an array of surfaces that reflect light and are projected onto the retina. Our perceptions of the visual world are not the geometrical characteristics of abstract space, but rather those of surfaces and edges. Thus, the spatial character of the visible world is given not by the objects in it, but by the surfaces of objects against a continuous background surface.

Second, there is always some variable in stimulation that corresponds directly with a property in the visual world. The stimulus-variable within the retinal image to which that property of visual space corresponds need only be a correlate of that property, not a copy.

Third, Gibson argued that in all environments the surfaces of ordinary, familiar things are of two types: a frontal surface, which is transverse to the line of sight and is best exemplified by the ground or terrain; and a longitudinal surface, which is parallel with the line of sight and is characteristic of objects. The perception of depth, distance, and the appearance of solidity is basically a problem of the perception of longitudinal surfaces.

Fourth, a general condition for the perception of a surface is the type of ordinal stimulation that yields texture. Gibson described ordinal stimulation as referring to the light energy on the retina, never to the objects from which light is reflected:

> let us assume that a stimulus is a type of variable physical energy falling within certain ranges of variation (the limits being called absolute thresholds) which excites a receptor or a set of receptors differentially. If it does not release physiological activity in the receptor-mechanism it is not a stimulus. As the energy varies successively, the energy varies concomitantly in some specific way. . . . We wish to extend this term to mean also a simultaneous variation over the set of receptors, or a differential excitation of different receptors, and the order of such a variation. For the extended meaning the term ordinal stimulation will be used (p. 63).

Just as the general condition for the perception of a surface was the ordinal type of stimulation that yields texture, so Gibson said that the general condition for the perception of an edge (and hence the perception of a bounded surface in the visual field) is the type of ordinal stimulation consisting of an abrupt transition. A change in brightness, for example, is such a retinal transition.

Fifth, the perception of an object in depth is reducible to the problem of the changing slant of a curved surface or the differing slants of a bent surface. In either case, Gibson noted that the problem was similar to how we see a longitudinal surface.

Finally, Gibson felt that the general condition for the perception of a longitudinal surface or a slanted surface is a kind of ordinal stimulation which he called a gradient. A gradient refers to nothing more than the increase or decrease of something along a given axis or dimension:

> we know from ordinary experience that the texture of different surfaces may vary from course to fine. . . . When the image of a single surface varies progressively in this way, it may be that the gradient of density is an adequate stimulus for the impression of continuous distance (p. 67).

In general, gradients are dependent upon outlines, retinal disparities, deformations when the observer moves, and shading—in short, they have all the function of stimulus correlates for the impression of distance on a surface (p. 76).

These six principles were at the heart of an enormous amount of research on spatial perception that Gibson reported in 1950 and subsequently in 1966 and 1979. As an indication of how his ideas changed over the course of many years, consider the fact that in 1979 he argued that our normal perception of pictures does not depend upon the ambiguous pictorial depth cues of traditional theory, but rather upon the structured light available to the perceiver that specifies not only places, things, people, and events but the perceiver's relation to that picture. Gibson (1979) wrote that

> the standard approach to vision begins with the eyes fixed and exposed to a momentary pattern of stimuli . . . The ecological approach to visual perception works from the opposite end. It begins with the flowing array of the observer who walks from one vista to another, moves around an object of interest, and can approach it for scrutiny, thus extracting the invariants that underlie the changing perspective structures and seeing the connections between hidden and unhidden surfaces. This approach next considers the fact of ambient awareness and explains it by the invariance of sliding samples of the 360 degree array. Only then is the awareness of a single scene considered, the surfaces seen with the head fixed and the array frozen. The classical puzzles that arise with this kind of vision are resolved by recognizing that the invariants are weaker and the ambiguities stronger when the point of observation is motionless. Finally, the kind of visual awareness obtained with eye fixed and the retina either briefly exposed or made to stay fixed is considered for what it is, a peculiar result of trying to make the eye work as if it were a camera at the end of a nerve cable. The visual system continues to operate at this photographic level, but the constraint imposed upon it are so severe that very little information can be picked up (pp. 303–304).

The above quotation is only one of many examples where we are pushed to rethink the very idea of what things look like and how we see them spatially.

As we have seen in this summary of the sensori-perceptual component, spatial perception entails the stipulation of how we perceive where things are, how far away they are, whether they are straight ahead or to the side, and which objects are next to each other. "Traditional" interpretations described the sensing of space in terms of cues, clues, or sensory attributes, and quantified our sensations and their cooperative differences in terms of our sensitivity to them. With regard to problems of solidity, distance, and direction, many traditional interpretations have employed the empty, abstract space of philosophers and physicists, and have characterized our awareness of extended space in Newtonian terms as an infinite, continuous, stationary, three-dimensional box without sides, a characterization that enables us to locate objects by their respective Cartesian coordinate values.

Gibson offered an exciting alternative to these interpretations, drawing upon our sensory experience of the visual field but emphasizing in particular our perceptual experiences in the visual world. Rather than argue that we gain meaning from cooperating senses or integrated cues, he argued for direct perception and assumed that every phenomenal aspect of the three-dimensional visual world, a world filled with familiar, ordinary objects, is represented by a corresponding aspect in the physical energy that affects our sensory receptors. Table 5.1 presents a comparison of traditional and Gibsonian ideas about perception in general.

A further examination of the works by Carr, Howard and Templeton, and Gibson would reveal a very large number of studies on sensory and perceptual awareness of space. Among topics we have mentioned are studies of upright and displaced vision, convergence and accommodation, monocular and binocular relations, accuracy of movement in different directions, eye movements and visual direction, the textural densities of different surfaces, auditory location, picture perception, accuracy of our perception of the apparent vertical, and the like. It is important to underline the fact that not only has our summary restricted itself to a very limited number of interpretations, but also it has only made reference to an enormous body of experimental literature about spatial perception. What needs to be added is that the problem of spatial perception, in all its different aspects, is still being investigated vigorously. The development of new models did not end with Gibson's reformulation, and it would be a serious mistake to assume that the problem of spatial perception has been solved, the claims of textbooks notwithstanding. Much the same can be said about the study of imagery, the next component to be discussed.

Imagery

As indicated in the introduction to this chapter, the role of imagery in thought has been the subject of speculation by philosophers for centuries. Thanks to the British empiricists, images came to be regarded as faint copies of movements and sensations associated with visual stimuli in memory, and obtained through repeated experiences with the external world. By the mid-1800s, however, imagery was also regarded by some as being a fundamental element of thought. Helmholtz discussed its relation to spatial awareness when he wrote:

TABLE 5.1. Gibson's Comparison of Traditional Versus Ideas About Perception

Traditional Theories	Gibson's Ideas
Perception is subjective, depends on contribution of observer; percept goes beyond stimuli and is superimposed on sensations.	"The visual worlds of different observers are more alike than they ought to be if the doctrine were the complete truth."
Perception is determined by processes in the brain that alter or supplement data which are available.	Perception is determined by the pick-up of data available to the perceiver directly.
Perception is a central response to peripheral reception activity.	Perception is a process of coordination between the observer and the relevant aspects of the environment.
Stimulation determines perception; stimulus precedes perception, which precedes cognition.	Perception does not depend on sensory impressions at all, but only on pick-up of stimulus information.
A difference between sensations and perceptions is assumed; sensations are bare and meaningless.	Environment is rich in "varied and complex stimulus information capable of giving rise to diverse perception through practice and exposure."
Sensory input must fuse with a stored image in memory.	Information does not have to be stored in memory, since it is always available.
Perception of visual or auditory stimulus is blocked through some special mechanism in the nervous system—filter theory of attention.	Perceiver ignores irrelevant information by simply not doing anything with it; selective attention.
Perception requires association and inference.	Perception requires differentiation and direct experience.
Meaning in the environment comes from associations, the imposition of meaning by "a little man in the brain."	The environment consists of substances, surfaces, places, objects, and events with meanings called affordances.
Perception is a process of matching to a representation in the head.	Perception is a process of extracting the invariants in stimulus information.
Order in the environment is pieced together by associating a cue with the event.	Order in the environment is perceived directly.
Living beings somehow construct a perceptually constant object from kaleidoscopic field of sensations.	Living beings are sensitive to the mathematically invariant properties in the stimulus flux that correspond to the physically constant object.
Improved perception involves the substitution of a new response to a stimulus.	Improved perception involves responding in a more discriminating way to a variable of stimulation.

Memory images of purely sensory impressions may . . . be used as elements of thought combinations without it being necessary, or even possible, to describe these in words. . . . Equipped with an awareness of the physical form of an object, we can clearly imagine all perspective images which we may expect upon viewing from this or that side, and we are immediately disturbed when such an image does not correspond to our expectations (Helmholtz, 1844, in Warren & Warren, 1968, pp. 252–254).

Although the study of imagery was systematically undertaken by psychologists of the Wurzburg school at the turn of this century, the reliance upon introspective reports raised serious questions about its role in thinking. As James (1912) put it:

"One can conjure up an image of a fire but one cannot be warmed by it: mental fire won't burn real sticks" (p. 96). Despite the fact that some psychometricians tied their descriptions of space to "the ability to obtain and to utilize visual spatial imagery" (e.g., El-Koussy, 1935), the study of imagery and of spatial ability was pursued as comparatively independent subjects for more than half of the present century.

The resurgence of interest in imagery in the 1950s has been the subject of many articles and books (e.g., Holt, 1964; Bruner, 1966; Sheehan, 1972; Neisser 1967; Richardson, 1969). It is of historical interest that a number of important contributions were all published during the same year (e.g., Paivio, 1971; Shepard & Metzler, 1971; Piaget & Inhelder, 1971; Segal, 1971). By the time Block (1981) wrote the introduction to his collection of articles about imagery, imagery had become "one of the hottest topics in cognitive science."

In the past fifteen years the literature has grown rapidly. Different theorists have characterized imagery as an ability, a skill, a mediating response, a style of thinking, a coding basis for representation, and a type of memory code. Imagery has also been characterized as having different properties. Anderson (1980), for example, listed six:

1. images are capable of representing continuously varying information;
2. they are capable of having operations performed on them that are analogues of spatial operations;
3. they are not tied to the visual modality, but seem to be part of a more general system for representing spatial and continuously varying information;
4. quantities such as size are harder to discriminate in images the smaller the quantities are;
5. images are more malleable and less crisp than pictures;
6. images of complex objects are segmented into pieces (p. 95).

Johnson-Laird (1983) maintains that efforts to describe images have divided psychologists into two opposing camps. In the first camp are Paivio (1971), Shepard (1978) and Kosslyn (1980), who argue that they are a distinct kind of mental representation. Johnson-Laird (1983) says that there is a consensus among these psychologists on four points:

1. The mental processes underlying the experience of an image are similar to those underlying the perception of an object or picture.
2. An image is a coherent and integrated representation of a scene or object from a particular viewpoint in which each perceptible element occurs only once with all such elements being simultaneously available and open to a perception-like process of scanning.
3. An image is amenable to apparently continuous mental transformations, such as rotations or expansions, in which intermediate states correspond to intermediate states (or views) of an actual object undergoing the corresponding physical transformation. Hence, a small change in the corresponds to a small change (of view) of the object.

4. Images represent objects. They are analogical in that structural relations between their parts correspond to the perceptible relations between the parts of the object represented (p. 147).

In the second camp are Baylor (1971), Pylyshyn (1973), and Palmer (1975), who argue that images are epiphenomenal, and that there is only a single underlying form of representation from which imagery may be generated.

Johnson-Laird also provides four points of consensus for the second camp. However, our concern at this point is not so much to get into this debate, as to describe in greater detail the models that Paivio, Shepard, and Kosslyn have proposed. We begin with a description of Paivio's dual-coding model.

Paivio's Model

Paivio (1971) maintained that long-term memory contained two qualitatively different but interconnected symbolic systems for coding and representing information:

> Images and verbal processes are viewed as alternative coding systems, or modes of symbolic representation, which are developmentally linked to experiences with concrete objects and events as well as with language. In a given situation, they may be relatively directly aroused in the sense than an object or an event is represented in memory as a perceptual image and a word as perceptual-motor trace, or they may be associatively aroused in the sense that an object elicits its verbal label (or images of other objects) and a word arouses implicit verbal associates or images of objects. In addition it is assumed that chains of symbolic transformations can occur involving either words or images, or both, and that these can serve a mediational function in perception, verbal learning, memory, and language (p. 8).

The functions of both verbal and imagery systems are coordinated in terms of a concrete-abstract dimension of stimulus meaning or task characteristics, with imaginal systems being more suited to concrete, and verbal systems more suited to abstract functioning. To demonstrate this dimension, Paivio showed that concrete words (those that name a pictorial object like a dog) are easier to recall than abstract words like "truth." He contended that concrete words can be memorized either imaginally or verbally, but abstract words can only be memorized verbally. Thus, we have *two* changes to remember a concrete word, by recalling either an image or a word, but only *one* chance to remember an abstract word.

A second postulated dimension of this model is the relative efficiency of imagery and verbal systems in the processing of symbolic information. For Paivio, imagery is a memory code for processing spatially parallel information; the verbal system serves to process sequential or serially-ordered information. Although it is tempting to regard imagery and verbal systems as dichotomous, Paivio insists that many everyday situations involve interaction, with imagery being

> a dynamic and transformable process without whose service "pure" verbal thinking might be less flexible and creative than has generally been asumed (p. 148).

With respect to individual differences, he argues that imagery is conceptually broader than the concept of spatial ability:

> Imagery can be regarded as a manifestation of the figural information that is in memory storage . . . In addition, however, imagery as a symbolic process is one class of determinant of cognitions – a cognitive tool that can be used in an anticipatory and creative fashion in such cognitive problems as (Guilford's 1967) figural transformations. In the latter sense, imagery is a construction based on stored information, and it is understandable that the storage and constructive utilization of imagery might constitute different abilities, which emerge empirically on different factors in the study of individual differences (p. 492).

Of interest is the fact that this suggestion subsequently has received support both in adults (DiVesta, Ingersoll & Sunshine, 1971) and in children (Forisha, 1975).

In a more recent work, Paivio (1986) emphasizes the functional nature of the two symbolic systems. Imagery, in particular, is characterized

> by its capacity for organizing multiple units of unrelated information quickly into synchronously-organized compounds, which can be efficiently redintegrated by a cue that gives access to a component. The retrieved information is simultaneously available in that it can be scanned and evaluated with relative freedom from sequential constraints, and it can be reorganized and transformed in various ways. Taken together, these characteristics imply that imagery contributes a richness of content and a flexibility in the processing of that content, so that diverse bits of information can be retrieved, compared, evaluated, transformed, and so on (p. 201).

Although Paivio has clearly broadened his characteristics of visual imagery, it is ironic that much of the research has been conducted using spatial ability tests as measures. Poltrock and Agnoli (1986), in their review of the literature concerning imagery and spatial abilities, noted that in some studies, scores from self-report questionnaires and spatial abilities were pooled to constitute a single index of imagery (e.g., Ernest, 1977; 1980; Ernest & Paivio, 1971; Paivio & Ernest, 1971). By contrast, in other studies spatial ability tests have served as the only measure of imagery (Paivio, 1978; Shaver, Pierson & Lang, 1974). Moreover, it has not been uncommon for investigators to employ spatial tests in extreme-group or median-split designs to assess the contribution of imagery to different spatial problems (Poltrock & Angoli, 1986, p. 265).

Overall, Paivio's dual-coding model has contributed to our thinking about the functional nature of imagery; about the relationship between imagery and language in memory and representation; and about the effects that different task characteristics and demands have upon our symbolic functioning.

Where Paivio (1971; 1986) has advanced a model to explain how imagery and verbal systems exist as separate and independent modes of information processing, Shepard (1978) has advanced one to explain how the formation and transformation of images contribute to our recognition of objects and to our solution of tasks in which we must compare one object with another. More specifically, where Paivio concentrated upon the relationship between imagery and language,

Shepard and his associates have looked at the relationship between the imagery and different aspects of perception.

Shepard's Model

The motivation for Shepard's model initially came from evidence that when people are asked to imagine that they are perceiving a rotating object, they appear to visualize it rotating in a continuous manner, in much the same way as it would actually rotate (Shepard & Metzler, 1971). In their study, Shepard and Metzler presented eight subjects with drawings of pairs of cubes. For each pair, the subject had to decide as quickly as possible whether or not the two figures depicted the same three-dimensional block shape. The block shapes were varied systematically with respect to the angle between the two figures, from 0 to 180 degrees. Subjects had to mentally rotate one figure in each pair until it aligned with the second figure. The further the figure had to be rotated, the longer it took to respond. Results showed a striking linear relationship between the angle of rotation and the time taken to make a comparison. Similar results were obtained when the pairs of cube structures were rotated in depth. These results were interpreted as evidence that when people make such comparisons, they perform some analogue of physical rotation itself.

Shepard (1978) rejected the notion that we have literal "pictures" in our heads (or "first-order" isomorphism) in favor of a more abstract or "second-order" isomorphism. As he noted,

> the proposed equivalence between perception and imagination implies a more abstract or "second-order" isomorphism in which the functional relations among objects as imagined must to some degree mirror the functional relations among objects actually perceived. Thus, to the extent that mental images can substitute for perceptual images, subjects should be able to answer questions about objects as well when those objects are merely imagined as when they are directly perceived (p. 131).

But what does it mean to speak of a purely mental process as being analogous to some external physical action? Shepard and Cooper (1982) argued that

> whatever neurophysiological events are taking place while one is merely imagining the external process in question, these events have much in common with the internal events that occur when one is actually perceiving the external process itself. As an empirically testable corollary, it implies that while one is in the course of imagining the external process, one passes through an ordered set of internal states of special relation to or readiness for the successive states of the external process (p. 206).

From studies of mental rotation and comparison, Shepard and his associates extended the chronometric paradigm to the study of various kinds of spatial transformations. As in the earlier studies, conclusions were based principally on the amount of time required to carry out a particular transformation and/or the amount of time required to match such a transformation against an external test stimulus.

For these later studies, the authors distinguished between three general cases of spatial transformations:

> (a) strictly rigid transformations, such as rotations, translations, and reflections or combinations of sequences of these, in which the whole object retains its rigid structure throughout; (b) what we might call "semirigid" transformations, such as foldings or joinings of discrete parts, in which each of those parts, though not the whole constellation of them, retains its separate rigid structure; (c) nonrigid transformations, such as plastic deformations, in which the topological structure of the object is globally preserved but in which every part or region may be subject to differential, locally affine stretchings and compressions (Shepard & Cooper, 1982, p. 244).

Moreover, they argued that, when making a comparison or establishing the equivalence of two objects related by some physical transformation, the brain is considered to pass through a trajectory of intermediate states that have a one-to-one correspondence to intermediate physical states, even when neither these physical states nor their retinal projections actually occur.

Shepard's model has prompted a large number of studies in which psychometrically established spatial tests or problems have been examined using an information-processing approach. This has contributed extensively to our knowledge about the processes involved in the solution of many spatial problems (e.g., the rotation of images, rotation of random two-dimensional shapes, a chronometric study of mental paper folding, and transformational studies of the representation of three-dimensional objects, to name a few). Moreover, as quoted above, Shepard has moved beyond a concern about imagery to one about different kinds of spatial transformations. Indeed, his distinction between three general kinds of transformation could serve as a useful preliminary classification of tasks that require an imaginal/spatial response.

Whereas Paivio and Shepard have both emphasized the functional nature of imagery, Kosslyn's (1980; 1983) model is more structural in that he claims that the internal structures of mechanisms used in the generation of images are similar to some of those used in perception. In particular, he has proposed a computer simulation model wherein mental images are assumed to function like pictures generated by a television set, in that they can be scanned, rotated, and otherwise transformed by various processing components.

Kosslyn's Model

Although Kosslyn's model is principally about imagery, it is also applicable to a wide range of problems involving psychological space. At the heart of it is the assumption that the mind generates and operates upon analogical representations, which preserve the spatial properties of visual stimuli. He distinguishes between structures and processes, arguing that there are two types of structures. The first is a visual buffer or form of short-term memory, which contains an array of "cells" that are activated whenever an image is formed. This array mimics a coordinate space, and supports data structures that provide information. According to Kosslyn, relations among activated portions of the visual buffer mirror the actual objects and their relationships, thus giving rise to the subjective impression that the image has a spatial character.

The second structure is the type of information stored in long-term memory. This may include both propositional information about the parts of objects and their relations, and information about an object's or an array of objects' literal appearances.

In addition to structures, Kosslyn postulates the existence of processes that operate upon the different structures. One process, which operates upon the visual buffers is called *Regenerate*. This process reactivates or refreshes the representation, which otherwise would fade with time. After all, if an image is to be operated upon, it must somehow be maintained. Other processes include *Rotate*, *Scan*, *Zoom*, and *Translate*. Each involves some transformation of the image, resulting in a modification of the representation in the visual buffer. Finally, Kosslyn says that *Find* and *Resolution* processes inspect and classify information in the representation.

The structure and processes of Kosslyn's model work together, and their interaction has been modeled within computer simulation programs. As Shwartz and Kosslyn (1982) describe it,

> The usual goal is to program the computer in such a way that its behavior emulates that of the human, presumably because both the mind and the program have the same functional states. . . . We use computer models of cognitive processes, one aimed at guiding an empirical research program rather than simply providing the bases for plausible accounts of data . . . We begin by conducting a series of experiments to delimit the class of acceptable theories . . . such that numerous choices about how to program the computer should be guided by the results of experiments (p. 69).

With respect to imagery, Kosslyn (1980) describes it as information generated in working memory as transitory data structures occurring in an analogue spatial medium. Kosslyn specifies four classes of image processes: *image generation*, *image inspection*, *image transformation*, and *image use*.

Image generation occurs when a person brings forth an image from more abstract representation in long-term memory. Image inspection results when one examines an image in order to answer questions about it. Image transformation is when one changes or operates upon an image: to solve a paper-folding task, for example, requires a succession of mental transformations. Image use occurs when an image is employed in the service of some other mental operation, such as geometry. An outline of the theory-relevant processes in Kosslyn's imagery model is presented in Table 5.2.

In a series of experiments, Kosslyn and his associates reported evidence for the perceptual-like functional properties of imagery. For example, they found that the relative size of objects is maintained in imagery; that it takes subjects longer to "see" features on animals that decrease in imagined size; that it takes longer to scan further distances across an imagined scene; and that when subjects are asked

TABLE 5.2. Outline of the Theory-Relevant Processes in Kosslyn's Model

Name	Type[1]	Input[2]	Operation	Output
PICTURE	P	r,θ file [size, location, orientation]	Maps point into surface matrix; mapping function may be adjusted to vary size, location, and/or orientation	Configuration of points depicting contents of an IMG file (produces new format; if mapping function adjusted, also produces new content)
FIND	C	Name of sought part	Looks up description; looks up procedures specified in description; executes procedures on surface matrix	Passes back Locate/Not Locate; if Locate, passes back Cartesian coordinates of part
PUT	P	Name of to-be-placed part	Looks up name of image file, location relation, and foundation part; looks up description of foundation part and relation; calls FIND to locate foundation part; adjusts mapping function; calls PICTURE	Part integrated into image (produces new content)
IMAGE	P	Name of to-be-imaged object(s) [size, location, orientation, level of detail]	Locates IMG file; calls PICTURE [if size, location, or orientation specified, adjusts mapping function; if detail required, searches for HASA entries, calls PUT]	Detailed or skeletal image at specified or default size, location, and/or orientation (produces a new content with different format, organization)
RESOLUTION	P	Surface image	Computes density of points in image	A number indicating dot density of image (produces new format)

TABLE 5.2. *Continued*

Name	Type[1]	Input[2]	Operation	Output
REGENERATE	A	Surface image	Works over surface matrix, refreshing most-faded parts first until all parts are refreshed	Image reactivated, with sharpness relations among parts altered (alters content)
LOOKFOR	P	Command to find a named part or property on an image	Calls REGENERATE; looks up description and size of part; calls RESOLUTION; if density not optimal, calls ZOOM or PAN; checks whether image overflows in direction of part, if so calls SCAN; calls FIND; if part not located, searches for relevant HASA entries, calls PUT to insert regions, calls FIND	Found/Not Found response
SCAN	A	Image, direction of required shift [rate]	Moves all points in surface matrix along vector; fills in new material at leading edge via inverse mapping function	Image repositioned (alters content)
ZOOM	A	Surface image, target resolution [rate]	Moves all points in surface matrix out from the center; fills in new material via inverse mapping function; calls RESOLUTION; calls PUT to insert new parts as resolution allows	Scale changes in image, higher resolution, and new parts (alters content)
PAN	A	Surface image, target resolution [rate]	Moves all points in surface matrix in from the center	Scale changes in image, lower resolution (alters content)

TABLE 5.2. *Continued*

Name	Type[1]	Input[2]	Operation	Output
ROTATE	A	Image, angle, and direction [rate]	Moves all points in bounded region in specified direction around a pivot	Reorients image (alters content).

[1]*A* indicates alteration transformations, which alter the initial data structure; *P* indicates production transformations that do not alter the initial data structure, but produce a new one from it; *C* indicates comparison operations that compare two data structures or parts thereof.
[2]Optional input is indicated in brackets.
Reproduced with permission from Kosslyn, S. (1980). *Mind and image*. Cambridge, Harvard University Press.

to imagine a scene, they keep the center of that scene's image in focus and "move" the image so that different parts appear at its center. Moreover, it was found that when subjects are asked first to imagine a distant object and then instructed to "walk around" it, smaller objects in the scene tend to "overflow" at nearer distances if they become larger upon approach.

Pellegrino, Alderton, and Shute (1984) summarized the different ways that Kosslyn's model is useful when discussing psychological space:

> First, it emphasizes the fact that the processing of visual-spatial information is composed of many basic processes that interact with information representations of varying detail and clarity. Second, tasks or performances can vary on several dimensions. One such dimension is the number of processes that must be executed to achieve a given result. Another dimension is the types of processes necessary to achieve that result. Third, individuals can vary in their performance depending upon how well they can perform certain processes and the extent to which those processes are necessary for solving different types of problems (p. 224).

As we have seen in this brief summary, within the past decade there has been a significant resurgence of interest in the functional and structural nature of imagery. Paivio (1971; 1986), Shepard (1978; 1981), and Kosslyn (1980) have contributed significantly to our understanding of the relationship between imagery and language, between imagery and perception, and between imagery and representation in general. In particular, they have helped us to understand more about how imagery relates to the processing of spatial information and the solution of various spatial tasks. It must be stressed that this summary is selective: other models have been advanced that also attempt to account for spatially represented information. Unfortunately, given the length constraints of this book, it is not possible to describe all of them.

We next turn to the third and last component of Berkeley's tri-part characterization of thought: association. Recall that philosophers of the empiricist persuasion and early experimental psychologists maintained that thought consists of sensory impressions, images, or faint copies of previous impressions in memory, and that we obtain our knowledge from the transformation, decomposition, and recombination of impressions and memories in associative thought. There is ongoing research on these ideas today.

Association

As indicated earlier, until the late 1950s a considerable amount of energy was directed by experimental psychologists into discovering how associations are formed and remembered. As summaries like Russell's (1956), Vinache's (1952), and Johnson's (1972) demonstrate, association or the study of "higher mental processes" was investigated primarily in terms of conceptualization or problem solving, with a gradual shift in interest from the development of single associations, to the competition among associations in different types of learning. Kendler (1964) and then Kessen (1966) expressed frustration with the lack of models to account for complex forms of thinking, and called for new metaphors of mind.

One response to this call was the elaboration of the computer as a metaphor of mind, and the rapid development of interest in information processing. By the mid-1970s, information processing had become a framework embracing a large number of research programs. The main division between theorists was between those whose goal was to develop computer programs that imitated or simulated human thought, and those who used the computer as a loose metaphor to help characterize the human processes of representing information and solving problems.

As Siegler and Richards (1985) note, theorists in the first group tend to

> start from a view of man as a manipulator or symbols. [Their] most basic goals are to describe the symbols that are represented (the representation) and to identify the ways in which they are manipulated (the processing). In trying to accomplish these goals, researchers must answer numerous questions about the nature of the information-processing system. Is information represented in linguistic form, in an imaginal form, or in some type of amodal propositional form? Is most processing conducted in parallel or serially? Is the overall system best conceived in terms of separate memory storage units such as immediate, short-term, or long-term memories, or in terms of a single system whose parts undergo varying degrees of activation at any one time (p. 920)?

By contrast, theorists in the second group attempt to identify the *sequence* of processes that people use to solve particular tasks or problems. There are three critical features to this approach. First, there must be some effort to describe the nature of the particular problem (e.g., visual, verbal, spatial, etc.). Next, there must be a method for determining which process or processes a subject employs on each trial or on most trials. And finally, there must be some means of identifying how the processes are combined in sequence in a solution strategy. As Carpenter and Just (1986) note, such methods as analyses of eye fixations, response times, and errors are particularly useful as convergent measures (p. 238).

In our descriptions so far, we have in fact already encountered a number of examples of models based upon an information-processing approach (e.g., Shepard & Cooper, 1982; Kosslyn, 1980). Rather than go over these earlier descriptions in an information-processing context, I will describe two additional models: namely, Egan's (1978; 1979) and Carpenter's and Just's (1986) models.

Egan's Model

Egan (1978) set out to answer the question: "Does the ability to perform quickly on a simple spatial operation predict the ability to perform accurately on more complicated spatial problems?" Judging from Zimmerman's (1954) study and other related psychometric literature, the two are not usually related.

In a series of studies, Egan (1976; 1979) attempted to relate items from psychometric measures of spatial ability to similar items recast within the parameters of information-processing tasks. He began by recasting items from the Guilford-Zimmerman Spatial Orientation test (varying the number of transformations to be performed on a mentally rotated clock) and from the Shepard-Metzler block rotation task. He then administered both the actual and the modified tasks to subjects, measuring response times and errors. Three measures of speed were used: total time, rotation rate (increase in response time as a function of the number of required transformations of the clock), and intercept (a measure of encoding, decision, and response times).

Egan (1978; 1979) found correlations among accuracy scores to be both high and positive, suggesting that the tests were in fact measuring what is commonly referred to as spatial ability. However, the rotation rates and the total response times were not correlated with accuracy scores. He found this pattern of results interesting, and argued that differences in speed and accuracy of repsonse may be unrelated because different processes are involved. To account for this possibility, he developed process models for the operations required for the solution of items from the spatial orientation and visualization tests. Figure 5.1 shows representations of these two models.

According to Egan (1979), efforts to relate measured spatial ability to the parameters of information-processing tasks could contribute to our understanding of psychological space in at least three ways:

> First, understanding tests in terms of mental operations suggests new interpretations for classical concepts used in psychological testing. These interpretations provide a conceptual, process-oriented framework to concepts that are most often treated in a purely actuarial, statistically-oriented fashion. Second, a deeper understanding of mental tests suggests new directions for work on individual differences. Of greatest importance is the reexamination of criterion scores in light of our understanding of predictors. Third, understanding tests in terms of mental processes may lead to improved methods for assessing abilities (p. 3).

Since Egan published his work, there have been other studies of a similar nature investigating individual differences at different ability levels for different

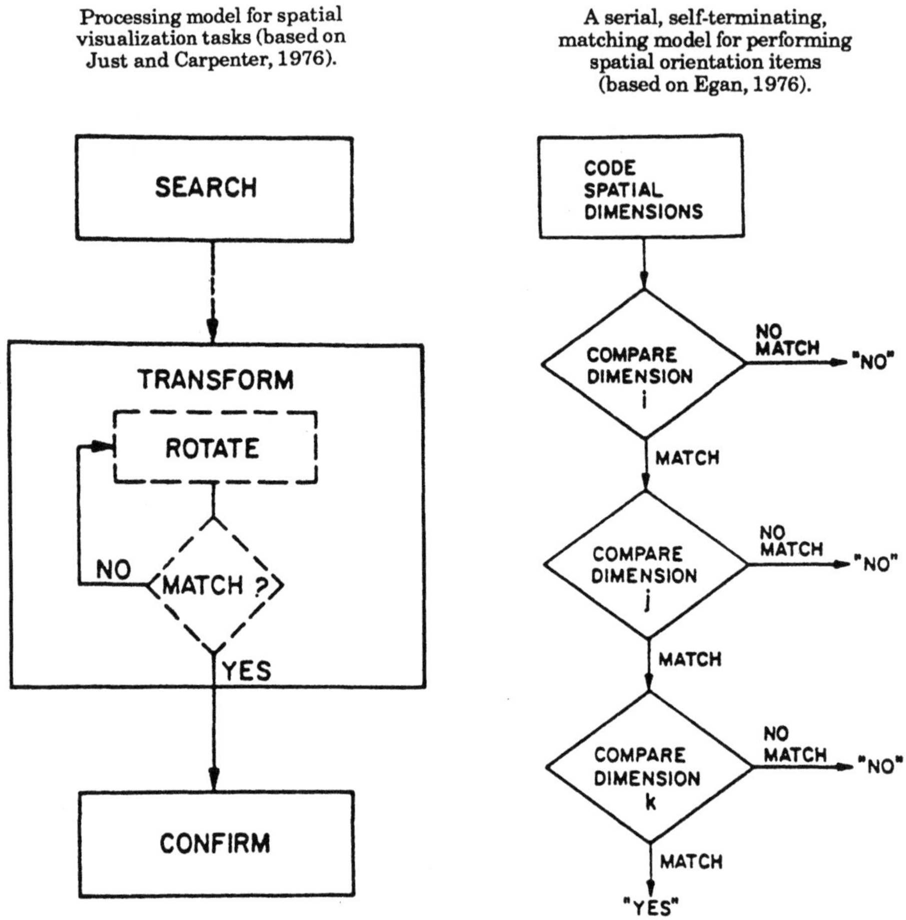

FIGURE 5.1. Examples of flowcharts of processes for spatial orientation and spatial visualization test items. (Reproduced with permission from Egan, D. [1979]. Testing based upon understanding: Implications from studies of spatial abilities. *Intelligence*, *3*, 1–15.)

kinds of spatial tasks with a variety of task characteristics (see, for example, Lohman, 1979; Just & Carpenter, 1985). Of interest is the model of spatial problem solving that Carpenter and Just (1986) recently proposed, which brings together much of this processing research.

The Carpenter and Just Model

This model is an attempt to synthesize findings from the psychometric and the information-processing literatures. It is a descriptive model in three parts: task characteristics, process characteristics, and strategy characteristics. Since many of the issues mentioned here have been discussed earlier, their presentation will be brief.

Carpenter and Just begin by discussing the nature of spatial representations. They suggest that individual differences may reside not so much in the *capacity* to represent, but in the spatial information to be represented. Common to the psychometric literature, in other words, is the possibility that individual differences occur in the *content* of spatial representations, as well as in how easily and accurately that content is maintained during processing.

A second task characteristic is the extent to which spatial representations are encoded with respect to a frame of reference. We respond to spatial information in several normally correlated frames, including the environmental context, gravity, and the up/down direction of the retina. For instance, tilting one's head in certain visual rotation tasks (Corballis & Beale, 1976) can affect speed of response, because such a behavior changes the encoded position of the stimulus relative to our frame of reference.

A third task characteristic is the extent to which a spatial representation preserves metric information. As Bundensen and Larson (1975) demonstrated, for example, response times increase with size disparity because it takes more time to make judgments about the properties of objects of different sizes.

A fourth characteristic is the extent to which spatial representations are related to perceptual processes (Finke, 1980). Brooks (1968) and Segal and Fusella (1970), for example, have shown that there are interference effects on spatial representations when two kinds of processes compete for the same resources. Carpenter and Just suggest that systematic studies of differences in the strength of such effects might serve as an index to whether a subject's representation of a given task or problem is indeed a spatial one.

With respect to process characteristics, Carpenter and Just describe generation, assessment, and transformational processes. In particular, they note that there are three ways in which spatial representations may be generated: by encoding the physical stimulus, by retrieving a previously constructed representation, and by constructing a new representation according to some non-iconic specification, like a verbal description. They observe that these three ways of generating representations are not independent of one another. For example, retrieval enters into both visual encoding and construction; and construction, in turn, may be required when generating a three-dimensional representation by encoding a two-dimensional picture. Moreover, different generating processes may entail different capacities during the execution of a task and, further, may interact with skill. Finally, they observe that

> whether the representation was originally retrieved, constructed or encoded could affect the precision and accuracy of the metric information, which features are represented, and how easily they are maintained during processing. Such differences could persist and even be exaggerated if, in the case of visual encoding, the stimulus remains available during the solution process (p. 242).

With respect to assessment procedures, the authors indicate that one way to explore the characteristics of a spatial representation is to ask subjects to describe its properties. Some spatial tests do not emphasize the manipulation of stimuli, but rather require the recognition and retention of features within them (see the classification of spatial tests in Figure 1.1). It is possible that performance differences on tasks requiring recognition and retention would provide information about the ease of generating representations, about the accuracy and detail of such representations, and about a subject's ability to preserve a representation while assessing its features. If Smith (1964) is correct in that what distinguishes "spatial" tests from other visual tests is the ability to perceive and to retain in mind a figure as an organized whole (p. 98), then assessment processes are clearly fundamental in the study of spatial representations.

Finally, with respect to transformational processes (processes that entail the mental manipulation of parts of a stimulus with respect to each other, to different frames of reference, to the observer's status), Carpenter and Just point out that in tasks requiring mental rotation or the taking of a different perspective, subjects have the choice of mentally rotating the stimulus, imagining a change in the perspective from which the object is viewed, or of constructing some orientation-free code for the object.

The authors maintain that the frame of reference used when encoding information about an object array acts as a cue to make it simpler to predict what kind of tasks will evoke different solution strategies:

> when the object is encoded with respect to a larger or stable frame of reference, the subject will manipulate the object's represented orientation. When the object is encoded as constituting part of the stable frame of reference, the subject will alter the representation of his own represented perspective. Finally, when the frame of reference is internal to the object, the code will be independent of the object's orientation (p. 245).

They go on to suggest that, in addition to frame of reference, investigations of transformational processes need to include such considerations as the process by which subjects decide the direction of transformation before it is executed, the direction or trajectory of change to the stimulus, the amount of information transformed during the change, and the processes that determine when the transformation is completed.

Finally, Carpenter and Just insist that their model of spatial problem solving requires attention to the subject's strategy selections for problem execution. This choice of strategy may be determined by two factors—the way in which the problem is presented, and the subject's ability to use a suggested form of representation. With respect to problem presentation, it has been pointed out elsewhere (Eliot & Smith, 1983) that considerable variation exists among apparently similar spatial tests with regard to differences in stimulus presentation, in the way stimulus items are organized and presented on a test page, in the quality of verbal instructions, in the response methods employed, and with regard to whether similar tests are administered within given time limits or not (p. 14).

There exists substantial evidence to support the claim that the way in which a spatial problem is presented affects performance differentially.

As Carpenter and Just also point out, the strategy selected for solving spatial problems may also be tied to a person's ability and level of skill. From the reports of subjects' own strategy choices, response times required to solve spatial tasks, and data on eye fixations, Just and Carpenter (1985) have developed simulation models of the processes that high and low spatial ability subjects use to solve Thurstone's (1938) Cube Comparison task. They found, among other results, that there are both qualitative and quantitative differences, with respect to the construct of spatial ability. Specifically, high ability spatial subjects are less likely to repeat a spatial transformation or regenerate a spatial representation, and appear to encode spatial information more accurately and in greater detail than low spatial ability subjects. As the authors conclude, individual differences may rest upon the distinction between subjects who are good at making spatial transformations as against those who are good at using non-transformational processes for solving the same problems (p. 250).

Lohman and Kyllonen (1983) examined differences in solution strategies for spatial tasks, and reported evidence for three kinds of intermediate relationships in a componential analysis of performance on a mental synthesis task. These relationships include instances where 1) ability level limits strategy selection; 2) strategy choice is unrelated to ability, but the effectiveness of implementing a strategy depends upon level of ability; and 3) ability both limits strategy choice and predicts performance within strategy groups. In their study, they tested three single-strategy models and two strategy-shift models for the encoding step in a task requiring the memorization of a stimulus figure. The single-strategy models were feature analysis (subjects stored each figure as a set of basic features), a figure decomposition model (breaking down complex figures into simpler ones), and a verbal labeling model (applying a descriptive label to a figure as a whole).

For the two strategy-shift approaches, one set of subjects were predicted to label some of the features of the stimulus figure and to decompose others, whichever was easier, while subjects in the second approach were predicted to label different proportions (25%, 50%, and 75%) of the figures, and to use the decomposition or feature-analysis strategy for the remainder.

Of thirty subjects in the study, Lohman and Kyllonen found that 11 were best fit by the feature-analysis model, 14 by the decomposition model, and 9 by the verbal labeling model (the remaining 6 could be described by more than one model). From their results, they concluded that subjects best fit by the decomposition or verbal labeling models were significantly quicker at obtaining a solution than those who were best fit by the feature-analysis model. Moreover, those best fit by the decomposition model obtained much higher scores on putative spatial tests.

In addition to providing more information about the relation between ability and strategy choice, Lohman and Kyllonen suggest that an important aspect of spatial ability may be skill in using prior figural knowledge when attempting to encode new figural stimuli. They also suggest that spatial ability may not simply entail the ability to generate visual images, but may also involve skill in reasoning with abstract visual images and with more concrete representations of these images.

This description of models and research using an information-processing approach may seem a long way from earlier experimental concerns with the formation and nature of associations, but in fact it is an extension of these earlier concerns. The terminology and the methodology have changed, but the preoccupation with the control and manipulation of behavior persists. Egan's (1978) process models for two spatial tests, based upon predicted differences in speed and accuracy of performance, represent some of the early studies using the information-processing approach; Lohman's and Kyllonen's (1983) study is a sampling of more recent research and Carpenter's and Just's (1986) model of spatial problem solving is a useful index of what has been gained in less than a decade.

Although much is now known about the processes and strategies that people use to solve problems, and although the examples of research given here have focused upon the processes used to solve items on psychometrically established measures of spatial ability, there is a danger in concluding that we have learned something about psychological space when in fact we have only learned more about how people solve so-called spatial tests. It is easy to confuse the convenience of tests that load upon a spatial factor with the definition of a problem domain. Just as some earlier theorists wondered whether single or competing associations represent the complexity of thought adequately, so some current theorists have reason to wonder whether our computer simulations or componential models for the processing of spatial information are necessarily generalizable to all nonverbal forms of representation or thought.

Issues

In this summary of the experimental approach, we began with a broad description of its characteristics and a brief history of the three components that make up the empiricist characterization of thought. These three components – sensory-perceptual space, imagery, and spatial processing – were then presented as parallel and ongoing histories, each with its representative models and a sampling of research. To conclude, we will turn to a consideration of the six general issues and some issues of specific concern to the experimental approach.

Although a considerable literature exists on the characterizations of thought, the issue of the emergence and decline of spatial perception, imagery, and spatial processing has not been a major concern of experimental researchers. Age-related changes in performance on tasks typically have been treated simply as one of several possible independent variables. As a result, while something has been learned about such changes in childhood and early adulthood, very little has been learned about changes in the latter half of life.

While it is tempting to cite representative studies from all components discussed in this chapter, I shall refer only to one example of research on age-related differences in spatial perception, and mention two examples on the ability to rotate images.

Wohlwill (1965) undertook a study of the relationship between age and judgments of perceived depth in photographic slides whose surfaces varied in density and regularity of texture. In the slides, a toy cow, horse, and fence were positioned in differently textured surfaces or fields at varying distances from each other. In one slide, for example, the cow was in the foreground, the horse in the background, and the fence somewhere in between. Children from the first, fourth, and eighth grades viewed these slides through an eyepiece that limited vision to a circular area containing the three objects, but cut out the borders of the screen so as to restrict information about the picture plane. Subjects made judgments about whether the cow or the horse was closer to the fence.

Wohlwill found that all age groups made consistent errors in judging the apparent midpoint (fence) to be further away than it actually was (over-constancy). Moreover, he found an overall decrease in constant error as texture density on the surface increased. Age differences were not an important variable, and no age and texture interaction was found. Wohlwill concluded that younger children do not depend more on the regularity of texture information than do older children when judging perceived distance in pictures. His findings are of particular interest considering how much more we have learned about pictorial space in the past twenty years (see Freeman, 1980; Freeman & Cox, 1985, for examples).

Somewhat more recently, Marmor and Zaback (1975), using the Shepard and Cooper (1982) technique for studying mental rotation, found that the ability to imagine rotational information was present in four-year-olds (Piaget and Inhelder notwithstanding). Specifically, four-year-olds showed the typical linear relation between degree of rotation and the amount of time needed to indicate whether two pictures were the same or mirror-images of each other. The slopes of functions were steeper than for adults, suggesting slower rates. Gaylord and Marsh (1975) found similar patterns for elderly adults when comparing their performances with those of the general adult population. Subsequent research has indicated that the rate of rotation approximately doubles between middle childhood and adulthood. Obviously, these studies provide only preliminary data, and we need more evidence from a variety of spatial problems before it is possible to generalize about the emergence and decline of spatial imagery or spatial processing.

Still more recently, Kail, Pellegrino, and Carter (1980) administered slides of the items from Thurstone's (1938) Primary Mental Abilities Spatial items to subjects from the third, fourth, and sixth grades and from college. They not only found additional evidence for developmental differences in the rate of rotation, but also found that, among young children, the additional time to encode and compare unfamiliar stimuli is the source of greatest individual variation, while among adults these processes plus the rate of mental rotation are equally large sources of individual differences (p. 170).

The Wohlwill, Marmor, and Kail et al. studies are just three examples of experimental research that have contributed to our knowledge about the emergence and decline of spatial perception, imagery, and processing. The issue of age-related changes in spatial behavior from the experimental approach is filled with researchable questions simply waiting to be investigated.

Something of the same can also be said for the issue of sex-related differences. Despite the fact that gender differences in cognitive functioning have stirred up quite a respectable literature, a cursory search of current textbooks on perception rarely indexes the topic (e.g., Hochberg, 1978; Rock, 1975; Sekuler & Blake, 1985). There have been more studies of sex-related differences in imagery and spatial processing, of which two examples will be described here.

Kail, Carter, and Pellegrino (1979) found that females generally had longer response times when making a comparison in a mental rotation task than did males. In fact, 30% of the females had longer response times than did all males. In addition to these group comparisons, they found greater variability in speed of response for females, with some performing like males while others formed a separate group with longer response times. This variability has been observed in a number of spatial tasks by other researchers, and warrants further attention.

The "water level" task, originally used by Piaget and Inhelder (1956), requires subjects to draw a line on a representation of a tilted cup indicating the appropriate water level. Since the original description of this task, it has been used by a large number of researchers because it frequently results in large sex-related differences (Thomas et al., 1973; Liben & Goldbeck, 1980).

Thomas (1973), for example, investigated the ability of hundreds of subjects to estimate the water level in a tilted jar. To succeed on this task, subjects have to understand the principle of horizontality (the water level remains parallel to the earth's surface regardless of the orientation of the container). Subjects from preschool through college were asked to predict the water level in a jar tilted in various orientations, and were scored in terms of the discrepancy between estimated and actual water level.

It was found that males in elementary school grasped the principle of horizontality, but that a significant number of college females did not. Males who were more accurate in their estimation would agree that "water is always level"; but females were more likely to agree that "water is level when the jar is upright, but is inclined when the jar is inclined" (Thomas et al., 1973).

More recently, Harris (1978) drew attention to the fact that if the task instructions emphasize analytic attention to task features, sex-related differences in performance on the water-level task are eliminated. When the instructions include the statement that "the water is in motion," males found the task more difficult than the standard task which presumes a container at rest. However, when the instructions included the statement that "the water is at rest," females found the task easier than the standard task. When the instructions included "the container is at rest" or "the container is in motion," results were equivalent to those for the standard task. These findings point to the fact that performance on the water-level task, like many other spatial tasks, may be more strongly affected by the algorithm for solution embedded in task instructions than has been supposed.

It is important to emphasize that most findings that provide evidence of sex-related differences on spatial tests are based upon trends rather than definite patterns of differences, with more variance in performance usually associated with individual differences rather than with sex-related differences per se. It will require more experimentation and less rhetoric to tease apart these differences with regard to ability level, knowledge base, skill experience, and social expectations, as they apply to a wide range of spatial tasks.

The third general issue, trainability, has received less attention from experimental psychologists than one might suppose. As indicated in earlier chapters, there are very few training or practice studies in the literataure that are undertaken for more than a brief interval of time, that employ a variety of related tasks, that report transfer and retention data, and do not "train" or provide specific practice or training for a particular criterion task. Even with a sensitivity to these and other constraints, it is difficult to find many well designed and well executed training studies of spatial perception, imagery, or spatial processing.

Kyllonen, Lohman and Snow (1984) attempted to train subjects' solution strategies for a paper-folding task by demonstrating alternative solution strategies and by systematically altering both item characteristics and mode of item presentation. They learned from a pilot study that three item characteristics accounted for most of the variance in item difficulty: the number of folds required, the number of obscured folds, the number of asymetric folds. When subjects in these pilot studies were asked to give retrospective descriptions of the strategies they had used, two strategies, one visualization and one analytic, were described most often. These two strategies were modeled on film, providing one half of the subjects in the main study with training in the visualization strategy, and the remainder with training in the analytic strategy.

The authors found that both training efforts were effective, depending upon the level of the subjects' ability and type of items presented. High spatial ability subjects performed best when they practiced solving items and were given feedback about their performance; subjects who were low in both spatial and verbal abilities performed best after receiving the visualization treatment. Kyllonen et al. concluded that strategic training for specific spatial tasks appears beneficial to some degree, especially when one capitalizes upon the cognitive strengths of particular subjects while compensation for their weaknesses in the face of specific task difficulties.

A continuing difficulty with studies that attempt to train spatial ability is the question of generalizability to other forms of perceiving or thinking about the surrounding world, and the use of knowledge about the "whereness" of things. The question actually has many facets: recall Bishop's (1980) earlier concern as to what constitutes a cognitive strength or weakness, and that training might have negative effects upon a person's overall cognitive functioning if it is directed toward changing a cognitive strength. Remember too that what has often been the target of practice or training programs are efforts to change an individual's use of sets of operations to solve particular problems, rather than increasing an individual's sensitivity to the layout of the surroundings in which one lives and moves.

The fourth issue, the role of imagery in spatial thinking, has already been dealt with at some length in this chapter. However, a large number of questions remain that have not been addressed. It is unclear, for example, how we recognize an image that is self-generated as against one that has perceptual reality. Also, while psychological research has advanced our knowledge about the functions and properties of visual imagery, it has told us little about the large-scale representations of those who lack visual imagery. Does the possession of verbal imagery lead to a fundamentally different kind of spatial representation? Hilgard put this question differently when he noted that, because the temptation is to assume that all senses can provide stimuli that produce images, and since images can be represented in each of our sense modalities, there must be some commonality to the different forms of imagery, which transcends the particular modality of each image. If this is so, does the visual imagery discussed in this chapter have properties in common with other sense modalities, to the extent that haptic or auditory imagery has a parallel role to visual imagery in spatial thinking? What happens to an individual's spatial functioning when he suffers the loss of imagery through traumatic injury or brain damage? (See Farah, 1984, for some thoughts about this question.)

To paraphrase Kosslyn (1980), it is likely that any detailed account of imagery at this early stage is almost certain to be incorrect. The controversy that continues to surround its study has not been so much with the end product itself as with the processes by which the end product is achieved. This distinction is an important one, not only for those interested in the relationship between imagery and spatial functioning, but for anyone interested in designing programs to foster or to enhance them.

With respect to the fifth issue, the types of processing required by spatial tasks, we have seen in this chapter how a large number of studies have reported that processing strategies vary according to 1) the subject's preference for a processing mode which may be related to ability level; 2) the interplay between stable and automatic processing routines and more labile attention-demanding processes; 3) the influence of task instructions that may force or suggest a particular strategy; 4) the effect of task demands such as speed, complexity, and the number of response alternatives that characterize a particular problem; and 5) individual differences in subjects' overall knowledge base and/or previous experience with spatial problems. We also obtained evidence that more complex tasks elicit more solution strategies than simple ones; that providing more time to complete tasks appears to encourage the use of less efficient strategies; and that there has been a pronounced tendency in the literature to view individual imagery differences in process-based rather than knowledge-based terms.

An interesting study of the strategies employed in the solution of different spatial tests was undertaken by Burden and Coulson (1981). These investigators argued that a common theme throughout the literature on spatial processing is that individuals differ with respect to their use of visual or verbal encoding, and with respect to their preference for responding to stimuli holistically or analytically. Accordingly, they began by developing an instrument for recording retrospective reports of subjects' strategies to solve spatial items, and then collected performance data from seven spatial tests along with these retrospective accounts. Evidence was sought that subjects represented test items visually, verbally, or in some form of mixed representation, and for whether they processed items using a sequential trial-and-error method of item solution (emphasizing the checking among parts of a stimulus figure), or by using a holistic method based upon treating the stimulus figure as a gestalt or single unit of information.

From these retrospective reports, Burden and Coulson found that it was possible to identify both the visual/verbal and the analytic/whole dimensions of solution strategies for a substantial proportion of subjects. They noted, for example, that subjects using a verbal strategy usually responded with words, phrases, and even complete sentences in their description of the test stimulus. Subjects would report that "they [the pieces for a paper formboard] are two triangles with long sides; together they make a diamond joined lengthwise." Subjects would insist that no "pictures-in-the-head" were involved, only words. As one subject put it, "I could see where they were joined, but I don't think I joined them as a picture in my head, I didn't need to." Whereas verbal descriptions most often took the form of "if X, then not Y, but Z," individuals who reported using visual strategies would report, "I can see that it cannot be X, but if I do this [some form of mental transformation], then I see that it must be Z."

With respect to analytic or holistic method preference, Burton and Coulson found that analytic or part-by-part processing was most commonly reported by subjects when they were asked to solve complex spatial tasks. The Embedded Figures Test, judged to be one of the more difficult of the seven tests administered, was most often solved using an analytic method. Typically, subjects would report that they selected a part of the simple stimulus figure and then searched for this part embedded in the more complex standard configuration. When they found the part, they reported using it as the basis for identifying the remainder of the simple figure.

Although Burden and Coulson were aware of the difficulties associated with retrospective reports of solution strategies (Ericsson & Simon, 1980), they were able to extract some interesting findings from their study. For example, they report that a visual-holistic strategy was more often reported for two-dimensional tasks, while a visual-analytic and a verbal-analytic strategy were most commonly used for more complex three-dimensional tasks. Of the four possible combinations, the verbal-holistic strategy was reported least often.

They also found that a majority of subjects drifted from a visual-holistic to a visual-analytic strategy between the first and second practice item on most spatial tests. Morevoer, the algorithm provided by the test instructions was the solution procedure used most often, even when instructions were modified to provide fewer clues. The sex of the subject strongly affected choice of solution strategies, and females employed a wider range of choices than males. Neither the time conditions associated with the different tests nor the instructional format, however, appeared to have a strong influence. Finally, it was found that most subjects using the analytic method also reported using mental imagery to solve spatial problems; more errors were made for items requiring a Same response than a Different response, with females making more errors on items requiring a Same response.

There is a long history of efforts to characterize solution processes in terms of dichotomous preferences: e.g., Walter's (1953) visualizers/verbalizers, French's (1965) systematic/analytic approach, Das's (1973) simultaneous/successive, Cooper and Podgorny's (1976) Type 1/Type 2, Willis's (1980) gestalt/analytic processing, and Burden's and Coulson's (1981) effort to examine two such dimensions or dichotomies. These studies have provided us with some interesting patterns with regard to the strategies used to solve different types of spatial tests. What would be of equal interest would be an extension of such studies by comparing processing strategies used by novice and skilled participants in activities that appear to require sets of operations for competent performance.

There have been a number of passing references in this chapter with respect to the sixth general issue, the effect of stimulus dimensionality upon spatial performance. Rosser (1980), for example, argued that mental rotation in two dimensions is easier and may reflect a different process than mental rotation in three-dimensions. Shepard and Cooper (1982), however, questioned this assertion, and reported finding no effect of stimulus dimensionality. They also suggested that the longer response times recorded for items presented three-dimensionally may reflect a dimensionality effect, or the fact that some subjects use inefficient strategies for processing more complicated three-dimensional stimuli. The point is that, at least for tests requiring mental rotation, differences in dimensionality may affect both response times and processing strategies, as Burden and Coulson (1981) suggested in a different context.

Another interesting example of the effect of stimulus dimensionality was reported by Wattawanaha (1977), who used a stratified sampling procedure to select 2,346 subjects (1201 boys and 1145 girls) from grades 7–9 in Australian schools. She then administered a wide range of paper-and-pencil spatial tests and, using the Rasch technique, developed the Monash Spatial Test with items that appeared to measure the same trait. When Wattawanaha and Clements (1983) subsequently administered the Monash Test to another large sample, they found that for 25 of the 75 comparisons for items, boys outperformed girls significantly. When these data were examined more closely, it was found that adolescent boys were more successful on items requiring a combination of three-dimensional thinking and the manipulation of mental images. There were not such sex-related differences on items requiring two-dimensional (figural) thinking and the use of static imagery.

Although many psychologists would agree that the ability to separate a figure or object from a larger configuration or background is basic to conceptualizations about perception, the extent to which we respond differently to particular figure-ground or object-background relationships amidst the flux of competing relationships is still not clearly understood. As indicated earlier, the issue of stimulus dimensionality goes beyond observer-object contexts to include observer-as-actor relationships in different environments.

With respect to issues that are more specific to the experimental approach, a variety of these have already been alluded to in the historical summaries of components and in the description of models within components. Concerning imagery, for example, there have been several references to the ongoing debate between those who argue that images are a distinct form of representation, and those who insist that they are only one expression of a single underlying form of representation. And concerning spatial processing, we have discussed how some theorists are concerned that more attention has been paid to differences in processes as a result of specific tasks, and less to the individual knowledge base and skill level that subjects bring to different tasks.

Other specific issues include questions about the relation between children's sensory integration and representation of extended space; the influence of different emotional states upon the formation and use of imagery; the extent to which information-processing models can be extended into studies of individual differences in sensory-perceptual as well as psychometrically-defined tasks; changes in the generation and use of imagery in elderly subjects' processing of spatial information; the issue of differences in social expectancies and cultural values in how extended space is perceived; or on the choice of solution strategies for particular activities or spatial tasks. Perhaps most important is the issue of the extent of retention and transfer of practiced or trained variables required in spatial processing. If this short list is any indication of researchable concerns, the experimental and cognitive psychologists still have much to contribute to our understanding of psychological space.

6
Themes and Recollections

In our consideration of the construct of psychological space, we have seen it characterized by both philosophers and psychologists in different ways. With respect to philosophers, there has been a gradual shift of interest from a concern about physical objects in the physical world in classical times, to an interest in mind and symbols which we associate with Hume, Kant, Berkeley, and other Enlightenment thinkers, to a more recent interest in the symbolic vehicles of thought. Within this progression, space has been characterized as a fundamental component of the world (Plato), as one of the ways we experience the physical world (Aristotle), as a necessary characteristic of "thinking matter" and "extended matter" (Descartes), and as a special component of our capacity to acquire knowledge—as an intuition rather than a concept—and a condition of our knowledge of the world as well (Kant). More recently, it has been characterized in terms of sensory cues, clues, attributes associated with objects, as a means for organizing information in terms of Cartesian coordinates, and as an encoding process for responding to spatial information in stimulation.

From this rich and diverse philosophical heritage, it is not surprising that psychologists have been selective in their choice of assumptions about psychological space. As we have seen, theorists have focused upon different sources of influence regarding the acquisition and use of spatial knowledge in their models. These different sources of influence can be described in rough terms as an internal-external continuum.

Among the models described, those advanced by Gesell, Werner, and Howard and Templeton are most concerned with the influence of bodily or muscular states upon our spatial perception and representation. Gesell (1949), it will be remembered, said that "the child comes by mind as he comes by his body, through the organizing principles of growth." Similarly, Werner (1964) argued that our perception is a reflection of the relation between proximal stimulation and ongoing organismic states. Whereas Gesell felt that at every stage of maturation the visual system undergoes changes that, in turn, serve to orient the ever-changing person, Werner contended that development depends on the degree to which an organism is organized, and the extent to which the separate components of this organization communicate with each other while functioning independently. From a different viewpoint, Howard and Templeton (1966) argue that adequate spatial behavior requires matching of the behavioral repertoire to the given constraints of the body and the world. All three theorists, in short, remind us that our spatial awareness of the external world, as well as our representation of that world, is constrained by ongoing and changing bodily influences.

The imposition of meaning upon sensory information and the directive function of perception is a second source of influence upon the acquisition and use of spatial knowledge. As we have seen, Baldwin (1915), Cassirer (1944), Werner (1946), Piaget (1956), Bruner (1964), and Olson and Bialystok (1983), among others, all posit the perceptual system as serving as an intermediary between our actions and our representations of space. Although most of these theorists have emphasized vision, presumably all sensory modalities contribute to our spatial understanding to some degree.

A third source of influence, probably the most commonly mentioned, is that of representation. In general, representation refers to the ways in which we remember, imagine, transform, and re-present spatial information to ourselves or communicate this information to others. The constituents of representation have not often been described in detail, and their relationship to different kinds of symbolic information as not always clearly articulated. Nevertheless, the capacity to use representations is an important source of influence upon spatial knowledge and skills in the models of Cassirer (1944), Werner (1946), Bruner (1966), Siegel and White (1975), Pick and Lockman (1981), Shepard and Cooper (1983), and Kosslyn (1980), among others.

A fourth source of influence is that of linguistic or symbolic systems. For Vygotsky (1962), remember, language is gradually internalized as private speech in the course of development and, as a consequence, the very structure of language becomes a vehicle for thought. Other theorists see symbolic manipulation as the crowning achievement of cognitive growth, that which distinguishes humans from all other species. Assuming the truth lies somewhere in between these viewpoints, it is evident that symbolic systems exercise a directive function for spatial thought in a number of ways. Examples are found in the models proposed by Werner (1946), Bruner (1964), and Olson and Bialystok (1983).

A fifth influence is the effect of task demand or environmental expectation upon spatial knowledge. Gestalt psychologists, recall, claimed that the structure of the external world was perceived and responded to as structure in thought. Similarly, Cassirer (1944) and Piaget (1956) both felt that the physical presence of objects affects our thought about them, as a consequence of our acting upon them. Although it may seem obvious that spatial tests have requirements that must be met in order to perform well, and although it may seem evident that different environments possess or are endowed with different response expectations (e.g., bedroom/classroom), not all models of psychological space take account of these requirements or expectations explicitly. Olson's (1970) notion of preformatory acts, Ittelson's (1973) distinction between object-centered and environmentally centered representation, and Liben's (1981) dimension of spatial products are three examples of this source of influence being explicitly incorporated into an account of spatial behavior.

The last source of influence upon spatial knowledge on this continuum is that of a value or meaning system. Olson and Bialystok (1983), for example, describe this source in terms of the purposes, goals, and intentions that are involved in our obtaining and using spatial information. Apart from the obvious differences in value and meaning systems in cross-cultural comparisons, more attention is needed with respect to the influence of these systems upon age-related and sex-related differences in the processing of spatial information.

This continuum is useful in that it shows us what a full description of psychological space would have to incorporate. Ideally, a comprehensive model would integrate all these sources of influence and their various combinations, and include a wide range of individual and group performances on diverse spatial tasks in a variety of environments. As indicated in the first chapter, such a comprehensive model is unlikely at present, because we lack a language system that would allow us to represent and communicate the complexity of different influences and their interactions for either a given moment or over the lifespan. The need for such a language system to characterize human thought is critical; the need for such a system to characterize psychological space is obvious.

The continuum is also interesting in that it parallels the continuum described by my graduate students in the first chapter. It is hoped that the discussions of various approaches to the study of psychological space have included descriptions that go beyond my students' first approximations at characterization.

Quite apart from the general issues discussed at the end of each chapter, a few additional issues cry out for attention. It should be evident, for example, that the primary focus of theories has been upon the relationship between self and objects and a secondary emphasis upon the relation between self and different environments. Within this focus, one value has been placed upon two-dimensional or figural space, another value upon three-dimensional or object space, and still a different value upon multi-dimensional or large-scale space. If these three kinds of space are understood differently at different times during the lifespan in different cultures (Weatherford, 1982; Mandler, 1984), then as Scheffler (1965) suggests, we need to consider what would constitute a standard by which various types of spatial ability, spatial knowledge, and spatial skill could be classified. Another way of saying this is to echo Wohlwill's (1973) remainder that spatial research needs both qualitative and quantitative efforts to be fully meaningful.

A different issue concerns the question of how spatial knowledge arises. As we have seen repeatedly, an assumption by many theorists has been that "to know requires one to act." Spatial knowledge, for Piaget, arises or is constructed from repeated actions that are gradually internalized and evolve as operations in schemata. Eventually, our knowledge serves to guide our action. As Miller et al. (1960) put it:

knowledge, action, and evaluation are essentially connected. The primary and pervasive significance of knowledge lies in its guidance of action knowing is for the sake of doing; action is obviously rooted in evaluation.

But although action serves as the motivational basis for knowing in many spatial models, other theorists suggest that alternative bases should be explored further. As Vygotsky (1962) and Olson and Bialystok (1983) both indicate, we may come to a significant portion of spatial knowledge through language. Similarly, as Allen (1978) points out, although Siegel and White (1975) and Pick (1973) both emphasize the role of action (or in this case the role of locomotion) in the development of large-scale representations, subjects in various studies have demonstrated route knowledge of a neighborhood without ever leaving the laboratory. While action and language are two obvious candidates as explanation for the acquisition and use of spatial knowledge, it is possible that other bases should also be considered.

This brief description of the continuum of sources of influence and mention of these issues are intended as a reminder of the need for a broader conceptualization of the construct of psychological space. In this book I have attempted to bring together models and evidence to support the contention that psychological space exists as an important and pervasive dimension of human thought and knowledge. Moreover, I have suggested repeatedly that the reformulated construct will be more than the sum of existing models and measures of spatial functioning, and that the construct itself will involve more than simply *our awareness of the relational distribution of things and our ability and skill to use this awareness to solve mental and physical problems*. Finally, I have tried to indicate in various ways that the reformulation of the construct should build upon what is known, but should also look to alternative philosophical and mathematical models for ways to characterize psychological space. In my judgment, the reformulation of the construct represents a challenge to scholars from many disciplines. The study of psychological space also constitutes a rich mine of researchable problems. Both the reformulation of the construct and the research associated with many facets should provide a significant increase in our understanding of how people perceive, represent, and use spatial information.

References

Ach, N. (1921). *Ueber die Begriffsbildung*. Bamburg: Buhner.

Acredolo, L.P. (1977). Developmental changes in the ability to coordinate perspectives of a large-scale space. *Developmental Psychology, 13*, 1–8.

Acredolo, L.P. (1978). Development of spatial orientation in infancy. *Developmental Psychology, 14*, 224–234.

Acredolo, L.P. (1979). Laboratory versus home: The effect of environment on the nine-month-old infant's choice of reference systems. *Developmental Psychology, 15*, 666–667.

Allen, G.L., Kirasic, K.C., Siegel, A.W., & Herman, J.F. (1979). Developmental issues in cognitive mapping: The selection and utilization of environmental landmarks. *Child Development, 50*, 1062–1070.

Allen, G.L. (1985). Strengthening weak links in the study of the development of macrospace cognition. In: Cohen, R. (Ed.). *The development of spatial cognition*. Hillsdale, NJ, Erlbaum Associates.

Allen, M.J. (1974). Sex differences in spatial problem-solving styles. *Perceptual and Motor Skills, 39*, 843–846.

Allport, F.H. (1955). *Theories of perception and the concept of structure*. New York: Wiley.

Anastasi, A. (1958). *Differential psychology* (3rd ed.). New York, Macmillan.

Anderson, G.V., Fruchter, B., Manual, H.T., & Worchel, P. (1954). *Survey of research on spatial factors*. Research Bulletin AFPTRC-TR-54-84. San Antonio, Texas, Lackland Airforce Base.

Anderson, J.R. & Bower, G.H. (1973). *Human associative memory*. Washington, DC, Winston.

Anderson, J.R. (1980). *Cognitive psychology and its implications*. New York, W.H. Freeman.

Annett, M. (1985). *Left, right, hand, and brain*. Hillsdale, NJ, Erlbaum Associates.

Aquinas, T. (1872–1880). *Opera omnia*. E. St. Fretlé & P. Maré (Eds.), Paris.

Aristotle. *Categories, 5*, 8–14.

Aristotle. *Le anima, 11*, 7; *111*, 1, 425.

Aristotle. (1831–1870). *Opera omnia*. I. Bekker (Ed.), Berlin.

Arnheim, R. (1969). *Visual thinking*. Berkeley, University of California Press.

Arnheim, R. (1974). *Art and visual perception*. Berkeley, University of California Press.

Averling, F. (1927). The relevance of imagery in the process of thinking. *British Journal of Psychology, 58*, 15–22.

Baeumker, C. (1908). *Witelo, ein Philosoph und Naturforscher des XIII.* Jahrhunderts, Münster, Aschendorff.

Badal, J. (1888). Contribution a l'etude des cecites psychiques: Alexie, agraphie, hemianopsie inferieure, trouble du sens de l'space. *Archives d' Ophtalmologie, 8,* 97–117.

Baldwin, J.M. (1915). *Mental development in the child and the race.* New York, Macmillan.

Baltes, P.B. & Willis, S.L. (1982). Plasticity and enhancement of intellectual functioning in old age. In: F.I.M. Craik & S.E. Trehub (Eds.). *Aging and cognitive processes.* New York, Plenum.

Barnes, E. (1893). A study of children's drawings. *Pediatric Seminars, 2,* 451–463.

Barragy, M. (1970). *Effects of varying object number and type of arrangement on children's ability to coordinate perspectives.* Unpublished dissertation, George Peabody College for Teachers, Nashville, Tennessee.

Barratt, E.S. (1953). An analysis of verbal reports of solving spatial problems as an aid in defining spatial factors. *Journal of Psychology, 36,* 17–25.

Barratt, P.E. (1953). Imagery and thinking. *Australian Journal of Psychology, 5,* 154–164.

Bartlett, F.C. (1932). *Remembering: an experimental and social study.* Cambridge, Cambridge University Press.

Baumrin, J.M. (1975). Aristotle's empirical nativism. *American Psychologist, 30(4),* 486–495.

Baylor, G.F. (1971). Programs and protocol analysis on a mental imagery task. *First International Joint Conference on Artificial Intelligence.*

Bayraktar, R. (1985). Cross-cultural analysis of drawing errors. In: N.H. Freeman & M.V. Cox (Eds.). *Visual order: The nature and development of pictorial representation.* New York, Cambridge University Press.

Beilin, H. (1971). The training and acquisitions of logical operations. In: M.F. Rosskopf, L.P. Steffe, & S. Tarback (Eds.). *Piagetian cognitive-development research and mathematical education.* Washington, DC: National Council of Teachers of Mathematics.

Bejar, I.I. (1985). Speculations on the future of test design. In: S. Embretson (Ed.). *Test design: Contributions from psychology, education, and psychometrics.* New York, Academic Press.

Bell, C. (1822). On the nervous circle which connects the voluntary muscles. Trans. Royal Society of London.

Bender, L. (1938). A visual motor gestalt test and its clinical use. *Research Monograph No. 3.* American Orthopsychiatric Association.

Benton, A.L. (1959). *Right-left discrimination and finger localization.* New York, Hoeper-Harper.

Berges, J. & Lezine, I. (1965). *Imitation of gestures.* London, Lavenham.

Berkeley, G. (1709). *A new theory of vision.* A.A. Luce & T.E. Jessop (Eds.), London.

Berkeley, G. (1710). *A treatise concerning the principles of human knowledge.* A.A. Luce & T.E. Jessop (Eds.), London.

Bierwisch, M. (1967). Some semantic universals of German adjectivals. *Foundations of Language, 3,* 1–36.

Binet, A. & Simon, T. (1905). Methodes nouvelles pour le diagnostic du niveau intellectuel des anormaux. *L'Annee' Psychologique, 11,* 191–244.

Bishop, A.J. (1980). Spatial abilities and mathematics—a review. *Educational Studies in Mathematics, 11,* 257–269.

Bishop, A.J. (1983). Space and geometry. In: R. Lesh & M. Landau (Eds.). *Acquisition of mathematical concepts and processes*. New York, Academic Press.

Blaut, J.M. & Stea, D. (1969). *Place learning*. Place Perception Research Reports. Worchester, Massachusetts, Clark University.

Block, N. (Ed.) (1981). *Imagery*. Cambridge, Massachusetts, MIT Press.

Bloom, B.S. & Broder, L. (1950). *Problem-solving processes of college students*. Chicago, University of Chicago Press.

Bock, R.D. & Kolakowski, D. (1973). Further evidence of sex-linked major gene influence on human spatial visualizing ability. *American Journal of Human Genetics, 25*, 1–14.

Boring, E.G. (1942). *Sensation and perception in the history of experimental psychology*. New York, Appleton-Century-Crofts.

Boring, E.G. (1946). The perception of objects. *American Journal of Psychology, 14*, 99–107.

Boring, E.G. (1950). *A history of experimental psychology*. New York, Appleton-Century-Crofts.

Boulding, K.E. (1973). *The image*. Ann Arbor, University of Michigan Press.

Bower, G.H. & Hilgard, E.R. (1983). *Theories of learning, 5th ed.* Englewood Cliffs, NJ: Prentice-Hall.

Bower, T.G. (1977). *A primer of infant development*. San Francisco, Freeman.

Bowers, H. (1931). Studies in visual imagery. *American Journal of Educational Psychology, 53*, 216–239.

Braine, L.G. & Lerner, C. (1980). Levels in the identifying of orientation by preschool children. *Journal of Experimental Child Psychology, 30(1)*, 171–185.

Brigham, C.C. (1932). *A study of error*. New York, College Entrance Examination Board.

Broad, C.D. (1933). The nature of a continuant. In: *Examination of McTaggart's philosophy (Vol. 1)*. Cambridge, Massachusetts, Cambridge University Press.

Bronfenbrenner, U. (1979). *The ecology of human development*. Cambridge, Harvard University Press.

Brooks, L.R. (1967). The suppression of reading by visualization. *Quarterly Journal of Experimental Psychology, 19*, 289–299.

Brooks, L.R. (1968). Spatial and verbal components in the act of recall. *Canadian Journal of Psychology, 22*, 349–368.

Broverman, D.M., Klaiber, E.L., Kobayashi, Y., & Vogel, N. (1968). Role of activation and inhibition in sex differences in cognitive abilities. *Psychological Review, 75*, 23–50.

Brown, W. & Stephenson, W. (1933). A test of a theory of two factors. *British Journal of Psychology, 23*, 352–370.

Bruner, J.S. (1960). *The process of education*. Cambridge, Harvard University Press.

Bruner, J.S. (1964). The course of cognitive growth. *American Psychologist, 19*, 1–15.

Bruner, J.S. (1966). *Studies in cognitive growth*. New York, Wiley.

Buffrey, A.H.H. & Gray, J.A. (1973). Sex differences in the development of spatial and linguistic skills. In: C. Oursted & D.C. Taylor (Eds.). *Gender differences: Their ontogeny and significance*. London, Churchill.

Bundensen, C. & Larsen, A. (1975). Visual transformation of size. *Journal of Experimental Psychology, 1(3)*, 214–220.

Burt, C. (1921). *Mental and scholastic tests*. London, P.S. King and Son.

Burt, C. (1949). The structure of mind: a review of the results of factor analysis. *British Journal of Educational Psychology, 19*, 176–199.

Burt, C.L. (1954). The differentiation of intellectual ability. *British Journal of Educational Psychology, 24*, 76–90.

Burden, L.D. & Coulson, S.A. (1981). *Processing spatial tasks*. Unpublished thesis, Victoria, Australia, Monash University.

Bussmann, H. (1981). *Mathematische föhigkeiten im didaktischen prozess*. Paderborn, Schöningh.

Butterworth, G.E. (1977). *Children's representation of the world*. New York, Plenum.

Buttimer, A. (1976). Grasping the dynamism of life world. *Annals of the Association of American Geographers, 66*, 277–292.

Caplan, P.J., MacPherson, G.M., & Tobin, P. (1985). Do sex-related differences in spatial abilities exist? *American Psychologist, 40*, 786–799.

Carpenter, P.A. & Just, M.A. (1986). Spatial ability: An information processing approach to psychometrics. In: R.J. Sternberg (Ed.). *Advances in the psychology of human intelligence. Vol. 3*. Hillsdale, NJ, Erlbaum Associates.

Carr, H.A. (1935). *An introduction to space perception*. New York, Hafner.

Carroll, J.B. (1982). The measurement of intelligence. In: R.J. Sternberg (Ed.). *Handbook of human intelligence*. Cambridge, Cambridge University Press.

Carroll, J.B. (1985). *Domains of cognitive ability*. A paper presented at the American Association for the Advancement of Science, Los Angeles, California.

Case, R. (1985). *Intellectual development: A systematic reinterpretation*. New York, Academic Press.

Cassirer, E. (1944). *An essay on man: An introduction to the philosophy of human culture*. New Haven, Yale University Press.

Cassirer, E. (1957). *The philosophy of symbolic forms, Volume 3*. New Haven, Yale University Press.

Cattell, J.McK. (1890). Mental tests and measurements. *Mind, 15*, 373–381.

Cattell, R.B. (1943). The measurement of adult intelligence. *Psychological Bulletin, 40*, 153–193.

Chomsky, N. (1957). *Syntactic structures*. S'Gravenhage, Netherlands, Mouton.

Churchill, B.D., Curtis, J.M., Coombes, C.H., & Hassell, T.M. (1942). The effect of engineer school training on the surface development test. *Educational and Psychological Measurement, 2*, 279–280.

Citrenbaum, C.H. (1975). *Spatial visualization, egocentricity, and field-dependence in criminal non-addicts, drug-free criminal addicts, and addicts maintained on methadone*. Unpublished dissertation, University of Maryland.

Clark, H.H. (1978). Space, time, semantics, and the child. In: T.E. Moore (Ed.). *Cognitive development and the acquisition of language*. New York, Academic Press.

Clark, H.H. & Clark, E.V. (1977). *Psychology and language: An introduction to psycholinguistics*. New York, Harcourt, Brace, Jovanovich.

Clements, M.A. (1983). The question of how spatial ability is defined and its relevance to mathematics education. *Zentralblatt fur Didaktik der Mathematik, 1(1)*, 8–20.

Cliff, N. (1966). Orthogonal rotation to congruence. *Psychometrika, 31*, 33–42.

Cohen, R. (1982). *Children's conceptions of spatial relations*. San Francisco, Jossey-Bass.

Cohen, R. (Ed.). (1985). *The development of spatial cognition*. Hillsdale, NJ, Erlbaum Associates.

Coie, J.D., Costanze, P.R., & Farnhill, D. (1973). Specific transitions in the development of spatial perspective-taking ability. *Developmental Psychology, 9(2)*, 167–177.

Cooper, L.A. & Podgorny, P. (1976). Mental transformations and visual comparison processes: effects of complexity and similarity. *Journal of Experimental Psychology, Human Perception and Performance, 2*, 503–514.

Cooper, L.A. & Regan, D.T. (1985). Attention, perception, and intelligence. In: H.J. Sternberg (Ed.). *Handbook of human intelligence*. Cambridge, Cambridge University Press.

Corballis, M.C. & Beale, I.L. (1976). *The psychology of left and right*. Hillsdale, NJ, Erlbaum Associates.

Cousins, J.H., Siegel, A.W., & Maxwell, S.E. (1983). Way finding and cognitive mapping in large-scale environments: A test of a developmental model. *Journal of Experimental Child Psychology, 35(1)*, 1–20.

Cratty, B.J. (1967). Perception of gradient and the veering tendency while walking without vision. *American Foundation for the Blind Research Bulletin, 14*, 31–51.

Cratty, B. & Sams, R. (1968). *The body-image of blind children*. New York, American Foundation for the Blind.

Cronbach, L.J. (1965). *Essentials of psychological testing*. New York, Harper and Row.

Darwin, C. (1859). *On the origin of the species*. London, John Murray.

Das, J.P., Kirby, J., & Jarman, R.F. (1979). *Simultaneous and successive cognitive processes*. New York, Academic Press.

Dawson, J.L.M. (1967). Cultural and physiological influences upon spatial-perceptual processes in West Africa. *International Journal of Psychology, 2(3)*, 171–185.

Dearborn, W.F. (1920). *Dearborn group tests of intelligence*. Philadelphia, Lippincott Company.

Deregowski, J.B. (1968). Difficulties in pictorial depth perception in Africa. *British Journal of Psychology, 59(3)*, 195–204.

Descartes, R. (1628). *Regulae ad directionem ingenii*. L. Gäbe & H. Springmeyer (Eds.), Hamburg, 1970.

Descartes, R. (1641). *Meditationes de prima philosophia*. A. Buchenau & L. Gäbe (Eds.), Hamburg, 1959.

DiVesta, F.J., Ingersoll, G., & Sunshine, P. (1971). Factor analysis of imagery tests. *Journal of Verbal Learning and Verbal Behavior, 10*, 471–479.

Dixon, R.A., Kramer, D.A., & Baltes, P.B. (1985). Intelligence: A life-span developmental perspective. In: B.B. Wolman (Ed.). *Handbook of intelligence*. New York, Wiley.

Dodge & Kirchway. (1913). *Teaching of geography in elementary schools*. Chicago, Rand McNally.

Dodwell, P. (1963). Children's understanding of spatial concepts. *Canadian Journal of Psychology, 17(1)*, 141–161.

Downs, R.M. & Stea, D. (1977). *Maps in minds*. New York, Hafner.

Downs, R.M. & Siegel, A.W. (1981). On mapping researchers mapping children mapping space. In: L. Liben, A. Patterson, & N. Newcombe (Eds.). *Spatial representation and behavior across the life-span*. New York, Academic Press.

Drever, J. & Collins, M. (1928). *Performance tests of intelligence: Series on non-linguistic tests for deaf and normal children*. Edinburgh, Oliver and Boyd.

Ebbinghaus, H. (1885). *Uber das Gedachtnis: Untersuchungen zur Eexperimentellen Psychologie*. Leipnig, Duncker and Humbolt.

Egan, D.E. (1976). *Accuracy and latency scores as measures of spatial information processing*. Pensacola, Florida, Naval Aerospace Medical Research Laboratory.

Egan, D.E. (1978). *Characterizing spatial ability: Different mental processes reflected in accuracy and latency scores*. Pensacola, Florida, Naval Aerospace Medical Research Laboratory.

Egan, D.E. (1979). Testing based upon understanding: implications from studies of spatial abilities. *Intelligence, 3,* 1–15.

Ekstrom, R.B., French, J.W., Harman, H.H., & Derman, D. (1976). *Kit of Factor-Referenced Cognitive Tests.* Princeton, NJ: Educational Testing Service.

Eliot, J. (1962). *Report on spatial relations unit.* Boston, Council for Public Schools.

Eliot, J. (1967). *The effects of age and training upon children's conceptualization of space.* NICHD Final Report HD 1648.

Eliot, J. & Smith, I.M. (1983). *International directory of spatial tests.* Windsor, England, NFER-Nelson.

Eliot, J. (1985). *DATASPACE: A topical search service for researchers interested in research and speculation about cognitive space.* College Park, Maryland, University of Maryland.

El-Koussy, A.A.H. (1935). Visual perception of space. *British Journal of Psychology: Monograph Supplement, 20,* 1–30.

Ellis, E. (1972). *The effects of pictorial mode on children's spatial visualization.* Unpublished dissertation, University of Maryland.

Ericsson, K.A. & Simon, H.A. (1980). Verbal reports as data. *Psychological Review, 87,* 215–251.

Ernest, C.H. & Paivio, A. (1971). Imagery and verbal associative latencies as a function of imagery ability. *Canadian Journal of Psychology, 25,* 83–90.

Ernest, C.H. (1977). Imagery ability and cognition: A critical review. *Journal of Mental Imagery, 1,* 181–216.

Ernest, C.H. (1980). Imagery ability and the identification of fragmented pictures and words. *Acta Psychologica, 44,* 51–57.

Farah, M.J. (1984). The neurological basis of mental imagery: A componential analysis. *Cognition, 18,* 245–272.

Faubian, R.W., Cleveland, E.A., & Hassell, T.W. (1942). Influence of training upon Mechanical Aptitude test scores. *Educational and Psychological Measurement, 2,* 91–94.

Fechner, G. (1966). *Elements of psychophysics.* Translated by Adler, H. New York, Holt, Reinhart, and Winston.

Fildes, L.G. & Myers, C.S. (1921). Left-handedness and the reversal of letters. *British Journal of Psychology, 3,* 273–278.

Fildes, L.G. (1923). Experiments on the problems of mirror-writing. *British Journal of Psychology, 3,* 57–67.

Fillmore, C.J. (1971). Toward a theory of deixes. In: D.D. Steinberg & L.A. Jakobovits (Eds.). *Semantics: An interdisciplinary reader in philosophy, linguistics, and psychology.* Cambridge, Cambridge University Press.

Finke, R.A. (1980). Levels of equivalence in imagery and perception. *Psychological Review, 87(2),* 113–132.

Finke, R.A. (1985). Theories relating mental imagery to perception. *Psychological Bulletin, 98(2),* 236–259.

Fishbein, H.D., Lewis, S., & Keiffer, K. (1972). Children's understanding of spatial relations: coordination of perspectives. *Developmental Psychology, 7,* 21–33.

Flavell, J.H. (1963). *Developmental psychology of Jean Piaget.* Princeton, NJ, Van Nostrand.

Flavell, J.H. (1970). Concept development. In: P.H. Mussen (Ed.). *Carmichael's manual of child psychology, vol. 1.* New York, Wiley.

Flavell, J.H. (1972). Analysis of cognitive-developmental sequences. *Genetic Psychology Monographs*, *86*, 279–350.

Flavell, J.H., Omanson, R.C., & Latham, C. (1978). Solving spatial perspective-taking problems by rule versus computation: A developmental study. *Developmental Psychology, 14(5)*, 462.

Flavell, J.H., Botkin, P.T., Fry, C.L., Wright, J.M., & Jarvis, P.E. (1968). *The development of role-taking and communication skills*. New York, Wiley.

Fonseca, S.H. & Kearl, B. (1960). *Comprehension of pictorial symbols: An experiment in rural Brazil*. Madison, University of Wisconsin College of Agriculture.

Forisha, B.D. (1975). Mental imagery and verbal processes. *Developmental Psychology, 11*, 259–267.

Fralley, J.S., Eliot, J., & Dayton, C.M. (1978). Further investigation of X-linked recessive gene hypothesis for spatial ability. *Perceptual and Motor Skills*, *47*, 1023–1029.

Freedman, S.J. (1968). *Neuropsychology of spatially-oriented behavior.* Homewood, Ill, Dorsey Press.

Freeman, F.N. (1916). Geography: Extension of experience through imagination. In: F.N. Freeman (Ed.). *The psychology of common branches*. Boston, Houghton-Mifflin.

Freeman, N.H. (1980). *Strategies of representation in young children: Analysis of spatial skills and drawing processes*. London, Academic Press.

Freeman, N.H. & Cox, M.V. (1985). *Visual order: The nature and development of pictorial representation*. Cambridge, Cambridge University Press.

French, J.W. (1951). Description of aptitude and achievement tests in terms of rotated factors. *Psychometric Monograph*, *5*.

French, J.W. (1954). *Manual for kit of selected tests for reference aptitude and achievement factors*. Princeton, NJ, Educational Testing Service.

French, J.W., Ekstrom, R.B., & Price, L.A. (1963). *Kit of reference tests for cognitive factors*. Princeton, NJ, Educational Testing Service.

French, J.W. (1965). Relationship of problem solving styles to factor composition of tests. *Educational and Psychological Measurement*, *25*, 9–28.

Fruchter, B. (1949). *The factorial content of the Airman classification battery.* San Antonio, Lackland Airforce Base.

Fruchter, B. (1952). Orthogonal and oblique solutions of a battery of aptitude, achievement, and background variables. *Educational and Psychological Measurement*, *12*, 20–38.

Frunchter, B. (1954). Measurement of spatial ability: history and background. *Educational and Psychological Measurement*, *14*, 387–395.

Furby, L. (1971). Role of spatial visualization in verbal problem solving. *Journal of Genetic Psychology, 85*, 149–150.

Galton, F. (1869). *Hereditary genius: An inquiry into its law and consequences*. New York, Appleton.

Galton, F. (1880). Statistics of mental imagery. *Mind*, *5*, 300–318.

Galton, F. (1883). *Inquiries into human faculty.* London, Macmillan.

Gardner, H. (1983). *Frames of mind*. New York, Basic Books.

Garrett, H.E. (1946). A developmental theory of intelligence. *American Psychologist*, *1*, 372–378.

Gaylord, S.A. & Marsh, G.R. (1975). Age differences in the speed of a spatial cognitive process. *Journal of Gerontology, 30*, 674–678.

Gesell, A. (1940). *The first five years of life*. New York, Harpers.

Gesell, A. & Ames, L.B. (1946). The development of directionality in drawing. *Journal of Genetic Psychology, 68*, 45–61.
Gesell, A., Ilg, F., & Bullis, G. (1949). *Vision: Its development in infant and child*. New York, Hoebler.
Ghent, L. (1961). Form and its orientation. *American Journal of Psychology, 74*, 177–190.
Gibson, J. (1950). *Perception of the visual world*. Boston, Houghton-Mifflin.
Gibson, J.J. (1966). *Senses considered as perceptual systems*. Boston, Houghton-Mifflin.
Gibson, J. (1979). *Ecological approach to visual perception*. Boston, Houghton-Mifflin.
Gibson, E.J., Gibson, J.J., Pick, A.D., & Osser, H.A. (1962). A developmental study of the discrimination of letter-like forms. *Journal of Comparative and Physiological Psychology, 55*, 897–906.
Gibson, E.J. (1969). *Principles of perceptual learning and development*. New York, Appleton-Century-Croft, 1969.
Glockner, H. (1968). *Die europäische Philosophie von den Anfängen bis zur Gegenwart*. Stuttgart, Reclam.
Goddard, H.H. (1910). A measuring scale for intelligence. *The Training School, 6*, 146–155.
Gombrich, E.H. (1960). *Art and illusion*. New York, Pantheon Books.
Goodenough, F.L. (1926). *Measurement of intelligence by drawings*. New York, Harcourt, Brace, and World.
Goodnow, J. (1977). *Children's drawing*. Cambridge, Harvard University Press.
Gordon, H. (1923). *Mental and scholastic tests among retarded children*. London, Board of Education.
Gosztonyi, A. (1976). *Der Raum*. Munich, Alber.
Gould, P. & White, R. (1974). *Mental maps*. Baltimore, Maryland, Penguin.
Gould, S.J. (1981). *The mismeasure of man*. New York, Norton.
Green, M.B. (1986). Superstrings. *Scientific American, 255*, 48–60.
Greenfield, P.M. (1976). Cross-cultural research and Piagetian theory: Paradox and progress. In: K.F. Riegel & J.A. Meacham (Eds.). *The developing individual in a changing world. vol. 1*. Chicago, Aldine.
Guay, R.B. & McDaniel, E.D. (1978). *Correlates of performance on spatial aptitude tests*. US Army Research Grant 19-77-G-0019. West Lafayette, Purdue University.
Guilford, J.P. (1959). Three faces of intellect. *American Psychologist, 3*, 3–11.
Guilford, J.P. (1967). *The nature of human intelligence*. New York, McGraw-Hill.
Guilford, J.P. (1985). The structure of intellect model. In: B. Wolman (Ed.). *Handbook of intelligence*. New York, Wiley.
Guilford, J.P. & Lacey, J.I. (1947). *Printed classification tests*. Washington, DC, Government Printing Office.
Guilford, J.P. & Zimmerman, E.S. (1948). The Guilford-Zimmerman aptitude survey. *Journal of Applied Psychology 32(1)*.
Guilford, J.P. & Hoepfner, R. (1971). *The analysis of intelligence*. New York, McGraw-Hill.
Gustafsson, J.E., Lindstrom, B., & Bjorck-Akesson, E. (1981). *A general model for the organization of cognitive abilities*. Report from the Department of Education, University of Götborg.
Gustafsson, J.E. (1982). *Visualization processes in learning as a function of method of presentation and individual differences*. Report from the Department of Education, University of Götborg.

Gustafsson, J.E. (1984). A unifying model for the structure of intellectual abilities. *Intelligence, 8*, 179–203.

Guttman, L. (1954). A new approach to factor analysis: the radex. In: P.F. Lzarfeld (Ed.). *Mathematical thinking in the social sciences*. Glencoe, Illinois, The Free Press.

Guttman, L. (1965). A faceted definition of intelligence. In: R.R. Eiferman (Ed.). *Scripta Hierosolymitana. vol. 14*. Jerusalem, Israel, Magnes.

Guttman, R. & Shoham, I. (1982). Structure of spatial ability items: a faceted analysis. *Perceptual and Motor Skills, 54(2)*, 487–493.

Haber, R.N. & Haber, R.B. (1964). Eidetic imagery 1: Frequency. *Perceptual and Motor Skills, 19*, 131–138.

Hagen, M.A. (1976). The influence of picture surface and station point on the ability to compensate for oblique view in pictorial perception. *Developmental Psychology, 12*, 57–63.

Hagen, M.A. (1980). *Perception of pictures*. New York: Academic Press.

Haggerty, M.E. (1920). *Intelligence examination: Delta series*.

Harris, A. (1957). Lateral dominance, directional confusion, and reading disability. *Journal of Psychology, 44*, 283–294.

Harris, D.B. (1963). *Drawings as measures of intellectual maturity*. New York, Harcourt, Brace, and World.

Harris, L.J. (1975). Neurophysiological factors in the development of spatial abilities. In: J. Eliot & N. Salkind (Eds.). *Children's spatial development*. Springfield, Illinois, Charles C Thomas.

Harris, L.J. (1978). Sex differences in spatial ability: possible environmental, genetic, and neurological factors. In: M. Kinsbourne (Ed.). *Hemispheric asymmetries of function*. New York, Columbia Press.

Hart, R.A. (1979). *Children's experience of place*. New York, Halstead.

Hart, R.A. & Moore, G.T. (1973). Development of spatial cognition. In: R.M. Downs & D. Stea (Eds.). *Image and environment*. Chicago, Aldine.

Hazen, N.L. (1983). Spatial orientation: a comparative approach. In: H.L. Pick & L.P. Acredolo (Eds.). *Spatial orientation: Theory, research, and application*. New York, Plenum.

Head, H. (1920). *Studies in neurology, vol. 2*. London, Oxford University Press.

Head, H. (1923). *Aphasia and kindred disorders of speech*. Cambridge, Cambridge University Press.

Healy, W. & Fernald, G. (1911). Tests for practical mental classification. *Psychological Monographs, 13*.

Heidbreder, E. (1933). *Seven psychologies*. New York, Century.

Hein, A.V. & Jeannerod, M. (1983). *Spatially oriented behavior*. New York, Springer-Verlag.

Helmholtz, H. (1925). *Handbook of physiological optics*. Translated by J.P.C. Southhall. New York, Optical Society of America.

Hilton, T. (1985). *National changes in spatial-visual ability from 1960 to 1980*. Princeton, NJ, Educational Testing Service.

Ho, K. (1974). *Exploratory study of spatial dimensionality across two ethnic groups at two age levels*. Unpublished dissertation, University of Maryland.

Hobbes, T. (1651). In: W. Molesworth (Ed.). *Leviathan, vol. 3*. London.

Hochberg, J.E. (1957). *Spatial representation: Theme 10*. Brussels, Proceedings of the International Congress of Psychology.

Hochberg, J.E. (1962). Nativism and empiricism in perception. In: L. Postman (Ed.). *Psychology in the making*. New York, Knopf.

Hochberg, J.E. (1978). *Perception*. Englewood Cliffs, NJ, Prentice-Hall.

Hoffman, K.I., Guilford, J.P., Hoepfner, R., & Doherty, W.J. (1968). *Factor analysis of figural-cognition and figural-evaluative abilities*. University of Southern California, Psychological Laboratory, Report No. 40.

Hollenberg, C.K. (1970). Functions of visual imagery in the learning and concept formation of children. *Child Development, 41*, 1003–1015.

Holly, C.D. & Dansereau, D.F. (1984). *Spatial learning strategies*. New York, Academic Press.

Holmes, A.C. (1963). *A study of understanding of visual symbols in Kenya*. London, Overseas Visual Aids Center.

Holt, R.R. (1963). Imagery: Return of the ostracized. *American Psychologist, 19*, 254–264.

Horn, J.L. (1972). Structure of intellect: primary abilities. In: R.H. Dreger (Ed.). *Multivariate personality research*. Baton Rouge, Louisiana, Claitor.

Horn, J.L. (1982). The aging of human abilities. In: B. Wolman (Ed.). *Handbook of intelligence*. New York, Wiley.

Horn, J.L. & Cattell, R.B. (1966). Age differences in primary mental abilities. *Journal of Gerontology, 21*, 210–220.

Horn, J.L. & Knapp, J.B. (1973). On the subjective character of empirical base of Guilford's structure of intellect model. *Psychological Bulletin, 80*, 23–42.

Houssaidas, L. & Brown, L.B. (1967). Coordination of perspectives by mentally defective children. *Journal of Genetic Psychology, 110*, 211–215.

Howard, I.P. & Templeton, W.R. (1966). *Human spatial orientation*. New York, Wiley.

Hudson, W. (1960). Pictorial depth perception in sub-cultural groups in Africa. *Journal of Social Psychology, 6*, 159–164.

Hudson, W. (1967). Study of pictorial perception among unacculturated groups. *International Journal of Psychology, 2*, 90–107.

Hughes, M. & Donaldson, M. (1979). Hiding games and coordination of viewpoints. *Educational Review, 31(2)*, 133–140.

Hull, C. (1928). *Aptitude testing*. Yonkers on Hudson, World Book.

Hume, D. (1739). *Treatise on human nature*. F.G. Selby-Bigge (Ed.), London, 1896.

Hume, D. (1748). *An enquiry concerning human understanding*. F.G. Selby-Bigge (Ed.), London, 1894.

Hunt, J.McV. (1961). *Intelligence and experience*. New York, Ronald.

Huttenlocher, J. (1968). Constructing spatial images: A strategy in reasoning. *Psychological Review, 75*, 550–560.

Huttenlocher, J. & Presson, C.C. (1973). Mental rotation and the perspective problem. *Cognitive Psychology, 4*, 277–299.

Huttenlocher, J. & Presson, C.C. (1979). Coding and transforming of spatial information. *Cognitive Psychology, 11*, 375–394.

Hyde, J.S. (1981). How large are cognitive gender differences? *American Psychologist, 36(8)*, 892–901.

Ilg, F.L. & Ames, M.B. (1964). *School readiness*. New York, Harper and Row.

Isard, W. (1979). *Spatial dynamics and optimal space time development*. New York, North Holland.

Ittelson, W.H. (1960). *Visual space perception*. New York, Springer-Verlag.

Ittelson, W.H. (1973). *Environment and cognition*. New York, Seminar.

James, W. (1912). *Essays in radical empiricism*. New York, Longmans Green.

Jammer, M. (1954). *Concepts of space*. Cambridge, Harvard University Press.

Jaynes, J. (1977). *Origins of consciousness in the breakdown of the bicameral mind.* Boston, Houghton-Mifflin.
Johnson, D.M. (1972). *A systematic introduction to the psychology of thinking.* New York, Harper and Row.
Johnson-Laird, P.N. (1983). *Mental models.* Cambridge, Harvard University Press.
Johnson, E.S. & Meade, A.C. (1987). Developmental patterns of spatial ability an early sex difference. *Child Development, 58,* in press.
Jöreskog, K.G. (1978). Structural analysis of covariance and correlation matrices. *Psychometrika, 43,* 443–477.
Jöreskog, K.G. (1981). *LISREL V. analysis of linear structural relationships by maximum likelihood and least squares methods.* Research report 81-8. University of Uppsala.
Just, M.A. & Carpenter, P.A. (1985). Cognitive coordinate systems: Accounts of mental rotation and individual differences in spatial ability. *Psychological Review, 92(2),* 137–172.
Kagan, J. & Kogan, N. (1970). Individual variation in cognitive processes. In: P. Mussen (Ed.). *Carmichael's manual of child psychology, Vol. 1.* New York, Wiley.
Kail, R.V., Carter, P., & Pellegrino, J.W. (1979). The locus of sex differences in spatial ability. *Perception and Psychophysics, 26(3),* 182–186.
Kail, R.V., Pellegrino, J.W., & Carter, P. (1980). Developmental changes in mental rotation. *Journal of Experimental Child Psychology, 29,* 102–116.
Kail, R.V., Stevenson, M.R., & Black, K.N. (1984). Absence of a sex difference in algorithms for spatial problem solving. *Intelligence, 8,* 37–46.
Kant, I. (1787). *Kritik der einen Vernunft.* Riga, Hartknoch.
Kelley, T.L. (1928). *Crossroads in the mind of man: A study of differentiable mental abilities.* Stanford, Stanford University Press.
Kendler, H.H. (1964). The concept of a concept. In: Melton (Ed.). *Categories of human learning.* New York, Academic Press.
Kennedy, J.M. (1974). *A psychology of picture perception.* San Francisco, Jossey-Bass.
Kephart, N.C. (1960). *The slow learner in the classroom.* Columbus, Ohio, Charles Merrill.
Kessen, W. (1966). Questions for a theory of cognitive development. In: Stevenson (Ed.). *Concept of development.* Monograph of the Society for Research in Child Development, *31,* 55–70.
Kielgast, K. (1972). Piaget's concept of spatial egocentrism: A re-evaluation. *Scandinavian Journal of Psychology, 12,* 179–191.
Kilbridge, P.L. & Robbins, M.C. (1968). Linear perspective, pictorial depth perception, and education among the Baganda. *Perceptual and Motor Skills, 27,* 601–602.
Kimball, M.M. (1981). Women and science: A critique of biological theories. *International Journal of Women's Studies, 4,* 318–330.
Kirasic, K.C. (1985). A roadmap to research for spatial cognition in the elderly adult. In: R. Cohen (Ed.). *The development of spatial cognition.* Hillsdale, NJ, Erlbaum Associates.
Köhler, W. (1947). *Gestalt psychology.* London, Liveright.
Kohs, S.C. (1923). *Intelligence measurement: A psychological and statistical study based upon the block designs test.* New York, Macmillan.
Kosslyn, S.M. (1983). *Ghosts in the mind's machine.* New York, Norton.
Kosslyn, S.M. (1980). *Image and mind.* Cambridge, Harvard University Press.
Kralovich, A.M. (1971). *The relationship of perceiving, recognizing, distinguishing, and*

relating to children's spatial visualization. Unpublished dissertation, University of Maryland.

Krumboltz, J.D. & Chrystal, R.E. (1960). Short-term practice effects in tests of spatial aptitude. *Personnel and Guidance Journal*, 384–391.

Kuhlman, C.K. (1960). *Visual imagery in children*. Unpublished dissertation, Radcliffe College.

Kyllonen, P.C., Lohman, D.F., & Snow, R.E. (1984). Effects of aptitude, strategy training, and task facets on spatial task performance. *Journal of Educational Psychology, 76(1)*, 130–145.

Langer, J. (1969). *Theories of development*. New York, Holt, Reinhart, and Winston.

Laurendeau, M. & Pinard, A. (1970). *The development of the concept of space in children*. New York, International Universities Press.

Lean, G.A. & Clements, M.A. (1981). *Spatial ability, visual ability, and mathematical performance*. Paupua, New Guinea, University of Technology.

Lean, G.A. (1981). Training of spatial abilities: a bibliography. *Educational Studies in Mathematics, 12(3)*, 267–299.

Leech, G.N. (1969). *Towards a semantic description of English*. Bloomington, Indiana, Indiana University Press.

Lesh, R. (1976). Apparent memory improvement over time. In: R. Lesh & D. Mierkiewicz (Eds.). *Recent research concerning the development of spatial and geometric concepts*. Columbus, Ohio, Ohio State University.

Liben, L.S. & Golbeck, S.L. (1980). Sex differences in performance on Piagetian tasks: differences in competence or performance? *Child Development, 51*, 594–597.

Liben, L.S. (1981). Spatial representation and behavior: Multiple perspectives. In: L.S. Liben, A. Patterson, & N. Newcombe (Eds.). *Spatial representation and behavior across the lifespan*. New York, Academic Press.

Liben, L.S. & Downs, R.M. (1986). *Children's production and comprehension of maps: Increasing graphic literacy*. Pennsylvania State University, Final Report NIE G-83-0025.

Linn, M.G. & Petersen, A.C. (1985). Emergence and characterization of sex differences in spatial ability: A meta-analysis. *Child Development, 56*, 1479–1498.

Locke, J. (1690). *Essays concerning human understanding*. A.C. Fraser (Ed.), Oxford, 1894.

Lohman, D.F. (1979). *Spatial ability: Review and re-analysis of the correlational literature*. Stanford, Stanford University Technical Report 8.

Lohman, D.F. (1985). *Dimensions of individual differences in spatial abilities*. Paper for NATO Advanced Study Institute in Cognition and Motivation, Athens.

Lohman, D.F. & Kyllonen, P.C. (1983). Individual differences in solution strategy on spatial tasks. In: R.F. Dillon & R.R. Schmeck (Eds.). *Individual differences in cognition, Vol. 1*. New York, Academic Press.

Lowenfeld, V. & Brittan, W. (1964). *Creative and mental growth*. New York, Macmillan.

Lunneborg, C.E. & Lunneborg, P.W. (1984). Contribution of sex-differentiated experiences to spatial and mechanical reasoning abilities. *Perceptual and Motor Skills, 59*, 107–113.

Luquet, G.H. (1927). *Le dessin enfantin*. Paris, Alcan.

Lynch, K. (1960). *The image of the city*. Cambridge, Harvard University Press.

Maccoby, E.E. & Bee, H. (1964). Some speculations concerning the lag between perceiving and performing. *Child Development, 36*, 367–378.

Maccoby, E.E. & Jacklyn, C.N. (1974). *Psychology of sex differences*. Stanford, Stanford University Press.

Magendie, F. (1822). Experiences sur les fonctions des racines des nerfs rachidiens. *Journal of Physiology and Experimental Pathology, 2*, 276–279.

Maitland, L. (1895). What children draw to please themselves. *Inland Educator, 1*, 87.

Mandler, J.M. (1982). Representation. In: J.H. Flavell & M. Markam (Eds.). *Handbook of child psychology, Vol. 3*. New York, Wiley.

Marmor, G.S. & Zaback, L.A. (1975). Development of genetic images: When does the child first represent movement in mental images? *Cognitive Psychology, 7*, 548–559.

Marmor, G.S. (1977). Mental rotation and number conservation: Are they related? *Developmental Psychology, 13*, 320–325.

Marshalek, B., Lohman, D.F., & Snow, R.E. (1983). The complexity continuum in the radex and hierarchical models of intelligence. *Intelligence, 7*, 107–128.

McCallum, D.I., Smith, I.M., & Eliot, J. (1979). A further investigation of components of mathematical ability. *Psychological Reports, 44*, 1127–1133.

McDonald, R.A. (1986). *Cognitive style, intelligence, age, and sex as related to performance on aerial photographic interpretation tasks*. Unpublished dissertation, University of Maryland.

McFarlane, M. (1925). A study of practical ability. *British Journal of Psychology, Monograph Supplement, 8*.

McGee, M.G. (1979). Human spatial abilities: Psychometric studies; Environmental, genetic, hormonal, and neurological influences. *Psychological Bulletin, 86(5)*, 889–918.

Michael, W.B., Zimmerman, W.S., & Guilford, J.P. (1950). An investigation into two hypotheses regarding the nature of the spatial relations and visualization factors. *Educational and Psychological Measurement, 10*, 187–213.

Michael, W.B., Guilford, J.P., Fruchter, B., & Zimmerman, W.S. (1957). Description of spatial visualization abilities. *Educational and Psychological Measurement, 17*, 185–199.

Mill, J. *Analysis of the phenomena of the human mind*.

Miller, J.W. & Miller, H.G. (1970). *A successful attempt to train children in coordination of projective space*. Nashville, Tenn, George Peabody College for Teachers., No. 70-03.

Miller, R.J. (1973). Cross-cultural research in the perception of pictorial materials. *Psychological Bulletin, 80(2)*, 135–150.

Miller, G. & Johnson-Laird, P.N. (1976). *Language and perception*. Cambridge, Harvard University Press.

Mitchelmore, M.C. (1976). *Perceptual development of Jamaican students*. Unpublished dissertation, Ohio State University.

Mitchelmore, M.C. (1978). Developmental stages in children's representation of regular solid figures. *Journal of Genetic Psychology, 133*, 229–240.

Mitchelmore, M.C. (1980). Predictions of developmental stages in the representation of regular space figures. *Journal for Research in Mathematics Education, 11*, 83–93.

Mitchelmore, M.C. (1985). Geometrical foundations of children's drawing. In: N.H. Freeman & M.V. Cox (Eds.). *Visual order: The nature and development of pictorial representation*. Cambridge, Cambridge University Press.

Mitchelmore, M.C. (1982). Variation in spatial abilities among industrialized countries. Cross-cultural. *Psychology Bulletin, 16(2)*, 2–3.

Moffie, D. (1940). *Non-verbal approach to Thurstone's primary mental abilities*. Unpublished dissertation, Pennsylvania State University.

Monroe, M. (1928). *Children who cannot read*. Chicago, University of Chicago Press.

Montangero, J. (1985). *Genetic epistemology: Yesterday and today.* New York, City University of New York.

Munn, N.L. (1954). Learning in children. In: L. Carmichael (Ed.). *Manual of child psychology.* New York, Wiley.

Murphy, G. (1948). *Historical introduction to psychology.* New York, Harcourt and Brace.

Myers, C.T. (1957). *Observations and opinions concerning spatial relations tests.* Princeton, NJ, Educational Testing Service, Research Memorandum RM-57-7.

Myers, C.T. (1958). *Observations of problem solving in spatial relations tests.* Princeton, NJ, Educational Testing Service, Research Memorandum RB-58-16.

Neale, J. (1966). Egocentrism in institutionalized and non-institutionalized children. *Child Development, 37,* 97–101.

Neisser, U. (1967). *Cognitive psychology.* New York, Appleton-Century-Crofts.

Neisser, U. (1976). *Cognition and reality.* San Francisco, Freeman.

Neumann, C. (1869). *Ueber die Prinzipien der Galitei-Newton'schen Theorie.*

Newcombe, N. (1982). Development of spatial cognition and cognitive development. *New Directions for Child Development, 15,* 65–82.

Newcombe, N. (1982). Sex-related differences in spatial ability: problems and gaps in current approaches. In: M. Potegal (Ed.). *Spatial abilities: Development and physiological foundations.* New York, Academic Press.

Newcombe, N., Bandura, M.M., & Taylor, D.G. (1983). Sex differences in spatial ability and spatial activities. *Sex Roles, 9(3),* 377–386.

Newcombe, N. (1985). Methods for studying spatial cognition. In: R. Cohen (Ed.). *The development of spatial cognition.* Hillsdale, NJ, Erlbaum Associates.

Newell, A., Shaw, J.C., & Simon, H.A. (1958). Elements of a theory of problem solving. *Psychological Review, 65,* 151–166.

Newton, I. (1779–1785). *Opera quae existant omnia.* S. Horsley (Ed.), London

O'Connor, J. (1943). *Structural visualization.* Boston, Human Engineering Laboratory.

O'Keefe, J. & Nadal, L. (1978). *Hippocampus as a cognitive map.* New York, Clarendon.

Ohta, R.J. (1983). Spatial orientation in the elderly: the current status of understanding. In: H.L. Pick & L.P. Acredolo (Eds.). *Spatial orientation: Theory, research and application.* New York, Plenum.

Olson, D.A. & Eliot, J. (1986). Relationships between experiences, processing style, and sex-related differences in performance on spatial tests. *Perceptual and Motor Skills, 62,* 447–460.

Olson, D.R. (1970). *Cognitive development.* New York, Academic Press.

Olson, D.R. (1975). On the relations between spatial and linguistic processes. In: J. Eliot & N. Salkind (Eds.). *Children's spatial development.* Springfield, Illinois, Charles C Thomas.

Olson, D.R. & Bialystok, E. (1983). *Spatial cognition: The structure and development of mental representations of spatial relations.* Hillsdale, NJ, Erlbaum Associates.

Orton, S.T. (1925). Word blindness' in school children. *Archives Neurological and Psychiatric, 14,* 581–615.

Paivio, A. (1971). *Imagery and verbal processing.* New York, Holt, Reinhart, and Winston.

Paivio, A. (1978). Comparisons of mental clocks. *Journal of Experimental Psychology, 4(1),* 61–71.

Paivio, A.(1986). *Mental representations: A dual coding approach.* New York, Oxford University Press.

Paivio, A. & Ernest, C.H. (1971). Imagery ability and visual perception of verbal and nonverbal stimuli. *Perception and Psychophysics, 10,* 429–432.

Palmer, S.E. (1975). Visual perception and world knowledge. In: D.A. Norman & D.E. Rumelhart (Eds.). *Explorations in cognition*. San Francisco, Freeman.

Papalia, D.E., Kennedy, E., & Sheehan, N. (1973). Conservation of space in non-institutionalized old people. *Journal of Psychology, 84(1)*, 75–79.

Pastore, N. (1971). *Selective history of theories of visual perception*. Oxford, Oxford University Press.

Paterson, D.G., Elliott, R.M., Anderson, L.D., Toops, H.A., & Heidbreder, E. (1930). *Minnesota mechanical ability tests*. Minneapolis, Minnesota, University of Minnesota Press.

Pattison, W.D. (1964). Four traditions in geography. *Journal of Geography, 63*, 211–216.

Pear, T.H. (1927). Relevance of visual imagery in the process of thinking. *British Journal of Psychology, 18*, 1–14.

Pellegrino, J.W., Alderton, D.L., & Regian, J.W. (1984). *Components of spatial ability*. Paper presented at NATO Advanced Institute in Cognition and Motivation, Athens, Greece.

Pellegrino, J.W., Alderton, D.L., & Shute, V.J. (1984). Understanding spatial ability. *Educational Psychologist, 19(4)*, 239–253.

Piaget, J. (1928). *Judgment and reasoning in the child*. London, Routledge and Kegan Paul.

Piaget, J. & Inhelder, B. (1956). *The child's conception of space*. London, Routledge and Kegan Paul.

Piaget, J., Inhelder, B., & Szeminska, A. (1960). *The child's conception of geometry*. New York, Basic Books.

Piaget, J. & Inhelder, B. (1971). *Mental imagery in the child*. New York, Basic Books.

Pick, H.L. (1972). *Mapping children – mapping space*. Honolulu, American Psychological Association.

Pick, H.L., Yonas, A, & Rieser, J. (1979). Spatial reference systems in perceptual development. In: M.H. Bornstein & W. Kessen (Eds.). *Psychological development from infancy*. Hillsdale, NJ, Erlbaum Associates.

Pick, H.L. & Lockman, J.J. (1981). From frames of reference to spatial representations. In: L.S. Liben, A.H. Patterson, & N. Newcombe (Eds.). *Spatial representation and behavior across the lifespan*. New York, Academic Press.

Pick, H.L. & Rieser, J.J. (1982). Children's cognitive mapping. In: M. Potegal (Ed.). *Spatial abilities: Development and physiological foundations*. New York, Academic Press.

Pick, H.L. & Acredolo, L.P. (1983). *Spatial orientation: Theory, research, and application*. New York, Plenum.

Pinter, R. & Paterson, D.G. (1917). *A scale of performance tests*. New York, Appleton.

Plato (1937). *Opera omnia*. J. Burnet (Ed.), Oxford.

Poltrock, S.E. & Brown, P. (1984). Individual differences in imagery and spatial ability. *Intelligence, 8(2)*, 93–138.

Poltrock, S.E. & Agnoli, F. (1986). Are spatial visualization ability and visual imagery ability equivalent? In: R.J. Sternberg (Ed.). *Advances in the psychology of human intelligence, vol. 3*. Hillsdale, NJ, Erlbaum Associates.

Porteus, S.D. (1915). Motor-intellectual tests for mental defectives. *Journal of Experimental Pedagogy, 3*, 127–135.

Potegal, M. (Ed.). (1982). *Spatial abilities: Development and physiological foundations*. New York, Academic Press.

Prauss, G. (1974). *Kant und das Problem der Dinge an sich*. Bonn, Bouvier.

Pribram, K.H. (1964). Neurological notes on the art of education. In: E.R. Hilgard (Ed.). *Theories of learning and instruction, Part 1*. Chicago, University of Chicago Press.

Price, L. & Eliot, J. (1975). Convergent and discriminant validities of two sets of measures of spatial orientation and visualization. *Educational and Psychological Measurement, 35(4)*, 975–977.

Pufall, P.B. & Shaw, R.E. (1973). Analysis of the development of children's spatial reference systems. *Cognitive Psychology, 5*, 151–175.

Pylyshyn, Z.W. (1973). What the mind's eye tells the mind's brain: a critique of mental imagery. *Psychological Bulletin, 80*, 1–24.

Pylyshyn, Z.W. (1981). The imagery debate: analogue media versus tacit knowledge. *Psychological Review, 88*, 16–45.

Quetelet, A. (1842). *A treatise on man and the development of his faculties*. Edinburgh, William and Robert Chambers.

Reese, H. & Overton, W. (1970). Models of development and theories of development. In:, L.R. Goulet & P.B. Baltes (Eds.). *Lifespan developmental psychology*. New York, Academic Press.

Reinert, G. (1970). Comparative factor analytic studies of intelligence throughout the human lifespan. In: L.R. Goulet & P.B. Baltes (Eds.). *Lifespan developmental psychology, research and theory*. New York, Academic Press.

Richardson, A. (1969). *Mental imagery*. New York, Springer-Verlag.

Rock, I. (1975). *Introduction to perception*. New York, Macmillan.

Roff, M.F. (1951). *Personnel selection and classification procedures: spatial tests, a factorial analysis*. USAF School of Aviation Medicine, Randolph Field, Texas. Project 21-29-002.

Rosser, R. (1980). *Acquisition of spatial concepts in relation to age and sex*. Final Report on Grant # NIE-6-79-0091 from the National Institute of Education, Department of Education. Tucson, University of Arizona.

Rowe, M. (1982). *Teaching in spatial skills requiring two- and three-dimensional thinking and different levels of internalization and the retention and transfer of these skills*. Unpublished dissertation, Monash University.

Royce, J.R. (1973). The conceptual framework for a multi-factor theory of individuality. In: J.R. Royce (Ed.). *Multivariate analysis and psychological theory*. New York, Academic Press.

Rubin, K.H., Attewell, P.W., Tierney, M.C., & Tumolo, P. (1973). Development of spatial egocentrism and conservation across the lifespan. *Developmental Psychology, 9*, 432.

Rushdoony, H.A. (1968). Child's ability to read maps: A summary of research. *Journal of Geography, 67*, 213–222.

Russell, D.H. (1956). *Children's thinking*. Boston, Ginn and Company.

Sachs, E. (1917). *Die fünf platonischen Körper*. Berlin, Weidmann.

Sauvy, J. & Sauvy, S. (1974). *The child's discovery of space*. Baltimore, Penguin.

Schaie, K.W. (1958). Rigidity-flexibility and intelligence: A cross-sectional study of the adult lifespan from 20 to 70. *Psychological Monographs, 72*, 462.

Schaie, K.W. & Strother, C.R. (1968). The cross-sequential study of age changes in cognitive behavior. *Psychological Bulletin, 70*, 671–680.

Schaie, K.W. (1970). A reinterpretation of age related changes in cognitive structure and functioning. In: L.R. Goulet & P.B. Baltes (Eds.). *Lifespan developmental psychology: Research and theory*. New York: Academic Press.

Schaie, K.W. (Ed.). (1983). *Longitudinal studies of adult psychological development*. New York, Guilford.

Scheffler, I. (1965). *Conditions of knowledge*. Chicago, Scott Foresman.

Segal, S.J. (Ed.). (1971). *Imagery: Current cognitive approaches*. New York.

Segal, S.J. & Fusella, V. (1970). Influence of imagined pictures and sounds upon detection of visual and auditory signals. *Journal of Experimental Psychology, 83*, 458–464.
Sekuler, R. & Blake, R. (1985). *Perception*. New York, Knopf.
Sharp, S.E. (1899). Individual psychology: A study in psychological method. *American Journal of Psychology, 10*, 329–391.
Shaver, P., Pierson, L., & Lang, S. (1974). Converging evidence for the functional significance of imagery in problem solving. *Cognition, 3*, 359–375.
Shaw, B. (1969). *Visual symbols survey.* London, Center for Educational Development Overseas.
Sheckels, M.P. & Eliot, J. (1983). Preference and solution patterns in mathematics performance. *Perceptual and Motor Skills, 57*, 811–816.
Sheehan, P.K. (1972). *Function and nature of imagery.* New York, Academic Press.
Shepard, R.N. (1975). Form, formation, and transformation of internal representations. In: R. Solso (Ed.). *Information processing and cognition: The Loyola symposium.* Hillsdale, NJ, Erlbaum Associates.
Shepard, R.N. (1978). The mental image. *American Psychologist, 33*, 125–137.
Shepard, R.N. & Metzler, J. (1971). Mental rotation of three-dimensional objects. *Science, 171*, 701–703.
Shepard, R.N. & Cooper, L.A. (1982). *Mental images and their transformations*. Cambridge, MIT Press.
Sheperd-Look, D.L. (1982). Sex differentiation and the development of sex roles. In: B. Wolman (Ed.). *Handbook of developmental psychology.* Englewood Cliffs, NJ, Prentice-Hall.
Sherman, J.A. (1967). Problem of sex differences in space perception and aspects of intellectual functioning. *Psychological Review, 74*, 290–299.
Sherman, J.A. (1979). Predicting mathematics performance in high school boys and girls. *Journal of Educational Psychology, 71*, 242–249.
Sherman, J.A. (1978). *Sex-related cognitive differences*. Springfield, Ill, Charles C Thomas.
Shwartz, S.P. & Kosslyn, S.M. (1982). A computer simulation approach to the study of mental imagery. In: J. Mehler, E.C.T. Walker, & M. Garratt (Eds.). *Perspectives on mental representation*. Hillsdale, NJ, Erlbaum Associates.
Siegel, A.W. & White, S.H. (1975). The development of spatial representations of large-scale environments. In: H. Reese (Ed.). *Advances in child development and behavior.* New York, Academic Press.
Siegler, R.S. & Richards, D.D. (1985). The development of intelligence. In: R.H. Sternberg (Ed.). *Handbook of human intelligence*. Cambridge, Cambridge University Press.
Sinka, D. & Shukla, P. (1974). Deprivation and development of skill for pictorial depth perception. *Journal of Cross-Cultural Psychology, 5*, 434–450.
Skinner, B.F. (1957). *Verbal behavior.* New York, Appleton-Century-Crofts.
Smith, I.M. (1964). *Spatial ability; Its educational and social significance*. London, University of London Press.
Smothergill, D.W. (1973). Accuracy and variability in the localizations of spatial targets at three age levels. *Developmental Psychology, 8*, 62–66.
Snow, R.E. (1978). Theory and method for research in aptitude processes. *Intelligence, 2*, 225–278.
Snow, R.E., Federico, P.A., & Montague, W.E. (Eds.). (1980). *Cognitive process analysis of aptitude: Vols. 1-2*. Hillsdale, NJ, Erlbaum Associates.
Snow, R.E. & Yalow, E. (1982). *Education and intelligence*. In: R.J. Sternberg (Ed.). *Handbook of human intelligence*. Cambridge, Cambridge University Press.

Spaulding, S. (1956). Communication potential of pictorial illustrations. *Audio-Visual Communication Review*, *4*, 31–46.

Spearman, C. (1904). "General intelligence" objectively determined and measured. *American Journal of Psychology*, *15*, 201–293.

Spearman, C. (1923). *The nature of intelligence and the principles of cognition*. London, Macmillan.

Spearman, C. (1927). *The abilities of man*. London, MacMillan.

Spearman, C. & Jones, L.W. (1950). *Human ability*. London, MacMillan.

Spencer, H. (1862). *First principles, 4th edition*. New York, Appleton.

Spiker, C.C. (1966). The concept of development: Relevant and irrelevant issues. *Monographs for the Society for Research in Child Development*, Series No. 107, *31(5)*, 40–54.

Stafford, R.E. (1961). Sex differences in spatial visualization as evidence of sex-linked inheritance. *Perceptual and Motor Skills*, *13*, 428.

Stafford, R.E. (1972). Negative relationships between ability to visualize space and grades in specific courses. *Journal of Learning Disabilities*, *5(1)*, 42–44.

Stea, D. & Blaut, J.M. (1973). Some preliminary observations on spatial learning in school children. In: R.M. Downs & D. Stea (Eds.). *Image and environment*. Chicago, Aldine.

Stern, W. (1909). The development of space perception in early childhood. *Z. Angew. Psychol*, *2*, 412.

Sternberg, R.J. (Ed.). (1985). *Handbook of human intelligence*. Cambridge, Cambridge University Press.

Stockbridge, F.P. & Trabue, M.R. (1921). *Measure your mind*. Garden City, NY, Doubleday and Doran.

Stratton, G. (1896). Some preliminary experiments on vision without inversion of the retinal image. *Psychological Review*, *3*, 611–617.

Stratton, G. (1899). Vision without inversion of retinal image. *Psychological Review*, *4*, 341–360.

Talmy, L. (1983). How language structures space. In: H.L. Pick & L.P. Acredolo (Eds.). *Spatial orientation: Theory, research, and application*. New York, Plenum.

Teller, P. (1969). Discussion and extension of Manfred Bierwisch's work on German adjectivals. *Foundation of Language*, *5*, 185–217.

Terman, L.M. (1921). Intelligence and its measurement: A symposium. *Journal of Educational Psychology*, *12*, 127–133.

Thomas, H., Jamison, E., & Hummel, D. (1973). Observation is insufficient for discovering that surface of still water is invariently horizontal. *Science*, *181*, 173–174.

Thomas, H. (1983). Familial correlational analysis, sex differences, and the x-linked gene hypothesis. *Psychological Bulletin*, *93(3)*, 427–440.

Thompson, G.B. (1975). Discrimination of mirror-image shapes by young children. *Journal of Experimental Child Psychology*, *19*, 165–176.

Thorndike, E.L. (1919). Group tests without language. Teachers College, Columbia University. (Also described in Thorndike, E.L. [1919]. A standardized group examination of intelligence independent of language. *Journal of Applied Psychology*, *3*, 13–32.)

Thorndike, E.L. (1924). The measurement of intelligence: present status. *Psychological Review*, *31*, 219–252.

Thurstone, L.L. (1931). Multiple factor analysis. *Psychological Review*, *38*, 406–427.

Thurstone, L.L. (1938). Primary mental abilities. *Psychometric Monographs*, *1*.

Thurstone, L.L. & Thurstone, T.G. (1941). Factor studies of intelligence. *Psychological Monographs*, *2*.

Thurstone, L.L. (1944). *Factorial studies of perception*. Chicago, University of Chicago.

Thurstone, G.T. & Thurstone, L.L. (1949). *Mechanical aptitude 11: Description of group tests*. Chicago, University of Chicago, No. 54.

Thurstone, L.L. (1951). *Mechanical aptitude 111: Analysis of group tests*. Chicago, University of Chicago, No. 55.

Trowbridge, C.C. (1913). Fundamental methods of orientation and imaginary maps. *Science, 38*, 888–897.

Tuddenham, R.D. (1962). Nataure and measurement of intelligence. In: L. Postman (Ed.). *Psychology in the making*. New York, Knopf.

Tukey, J.W. (1977). *Exploratory data analysis*. Reading, Massachusetts, Addison-Wesley.

Twyman, A. & Bishop, A.J. (1972). *Imagery and mathematical ability*. London, Chelsea College.

Tyler, L.E. (1976). The intelligence we test—an evolving concept. In: L.B. Resnick (Ed.). *The nature of intelligence*. Hillsdale, NJ, Erlbaum Associates.

van der Horst, L. (1950). Affect, expression, and symbolic function in the drawing of children. In: M.L. Reymert (Ed.). *Feelings and emotions*. New York, McGraw-Hill.

Vandenberg, S.G. (1969). A twin study of spatial ability. *Multivariate Behavioral Research, 4*, 273–294.

Vandenberg, S.G. (1975). Sources of variance in performance of spatial tests. In: J. Eliot & N. Salkind (Eds.). *Children's spatial development*. Springfield, Illinois, Charles C Thomas.

Vernon, P.E. (1940). *The measurement of abilities*. London, University of London Press.

Vernon, P.E. (1947). Research on personnel selection in the Royal Navy and the British Army. *American Psychologist, 2*, 52–63.

Vernon, P.E. (1950). *The structure of human abilities*. London, Methuen.

Vernon, P.E. (1969). *Intelligence and cultural environment*. London, Methuen.

Vinache, E.W. (1952). *The psychology of thinking*. New York, McGraw-Hill.

Vogel, J.M. (1980). Limitations on children's short-term memory for left-right orientation. *Journal of Experimental Child Psychology, 30(3)*, 473–495.

von Senden, M. (1960). *Space and sight*. New York, Free Press.

Vygotsky, L.S. (1962). *Thought and language*. New York, Wiley.

Vygotsky, L.S. (1978). *Mind in society*. Cambridge, Harvard University Press.

Waber, D.P. (1977). Sex differences in mental abilities, hemispheric lateralization, and the rate of physical growth at adolescence. *Developmental Psychology, 13*, 29–38.

Waber, D.P. (1979). Cognitive abilities and sex-related variations in the maturation of cortical functions. In: M.A. Zittig & A.C. Petersen (Eds.). *Sex-related differences in cognitive functioning*. New York, Academic Press.

Wagner, H. (1967). *Philosophie und Reflexion*, 2nd ed. München, Reinhart.

Wallon, H. (1945). *Les origines de la pensee chez l'enfant*. Paris.

Walsh, D.A., Krauss, I.K., & Regnier, V.A. (1981). Spatial ability, environmental knowledge, and environmental use: The elderly. In: L.S. Liben, A.H. Paterson, & N. Newcombe (Eds.). *Spatial representation and behavior across the lifespan*. New York, Academic Press.

Walter, W.G. (1953). *The living brain*. London, Duckworth.

Wapner, S. & Werner, H (1957). *Perceptual development*. Worchester, Massachusetts, Clark University Press.

Wapner, S. (1969). Organismic-developmental theory: some applications to cognition.

In: P.H. Mussen, P. Langer, & M. Covington (Eds.). *Trends and issues in developmental psychology.* New York, Holt, & Winston.
Wapner, S.A., Cirillo, L., & Baker, A.H. (1971). Some aspects of the development of space perception. In: J.P. Hill (Ed.). *Minnesota symposium on child development.* Minneapolis, University of Minnesota Press.
Wapner, S.A. & Cirillo, L. (1968). Imitation of model's hand movements: Age changes in the transposition of left-right relations. *Child Development, 39,* 887–894.
Warren, R.M. & Warren, R.P. (1968). *Helmholtz on perception.* New York, Wiley.
Wattanawana, N. (1977). *Spatial ability and sex differences in performance on spatial tasks.* Unpublished dissertation, Monash University.
Wattanawana, N. & Clements, M.A. (1982). Qualitative aspects of sex-related differences in performance on pencil-and-paper spatial questions in grades 7–9. *Journal of Educational Psychology, 74,* 878–887.
Weatherford, D.L. (1981). Spatial cognition as a function of size and scale of the environment. *New Directions for Child Development, 15,* 5–18.
Webster's seventh new collegiate dictionary. (1963). Springfield, Massachusetts, G and C Merriam.
Wechsler, D. (1950). Cognitive, conative, and non-intellective intelligence. *American Psychologist, 5,* 78–83.
Wellman, H.M. (1985). *Children's searching: The development of search skill and spatial representation.* Hillsdale, NJ, Erlbaum Associates.
Werdelin, I. (1961). *Geometrical ability and the space factor.* Lund, Gleerup.
Werner, H. (1948). *Comparative psychology of mental development.* New York, Harper and Row.
Wertheimer, M. (1923). Untersuchugen zur lehre von der gestalt. *Psychologie Forsche, 4,* 301–350.
Wherry, R.J. (1959). Hierarchical factor solutions without rotation. *Psychometrika, 24,* 45–51.
Whipple, G.M. (1915). *Manual of mental and physical tests.* Baltimore, Warwick and York.
White, S.H. (1970). The learning theory and child psychology. In: P.H. Mussen (Ed.). *Carmichael's manual of child psychology, Vol. 1.* New York, Wiley.
Whitehouse, D., Dayton, C.M., & Eliot, J. (1980). A left-right identification scale for clinical use. *Developmental and Behavioral Pediatrics, 1(3),* 118–121.
Willis, G. (1980). Spatial ability: Some neurological research. *MERGA, 2,* 187–197.
Willis, S.L., Blieszner, R., & Baltes, P.B. (1981). Intellectual training research in aging: Modification of performance on the fluid ability of figural relations. *Journal of Educational Psychology, 73,* 41–50.
Windelband, W. & Heimsoeth, H. (1957). *Lehrbuch der Geschichte der Philosophie.* 15th ed. Tübingen, Mohr.
Winter, W. (1963). The perception of safety posters by Bantu industrial workers. *Psychologica Africana, 10,* 127–135.
Wissler, C. (1904). The correlation of mental and physical tests. *Psychological Review Monographs Supplement, 3,* 4.
Wohlwill, J.F. (1965). Texture of the stimulus field and age as variables in the perception of relative distance in photographic slides. *Journal of Experimental Child Psychology, 2,* 163–177.
Wohlwill, J.E. (1973). *The study of behavioral development.* New York, Academic Press.

Wolman, B. (1985). *Handbook of intelligence*. New York, Wiley.

Woodring, P. (1939). *An investigation of directional orientation*. Unpublished dissertation, Ohio University.

Woodworth, R.S. & Schlosberg, H. (1954). *Experimental psychology*. New York, Holt, Reinhart, and Winston.

Woolley, H.T. (1915). A new scale of mental and physical measurements for adolescents. *Journal of Experimental Psychology*, *66*, 521–550.

Yen, W.M. (1975). Sex-linked major-gene influences on selected types of spatial performance. *Behavior Genetics*, *5*, 35–41.

Zimmerman, W.S. (1953). A revised orthogonal rotation solution for Thurstone's original primary mental abilities test battery. *Psychometrika*, *18*, 77–93.

Zimmerman, W.S. (1954). The influence of item complexity upon the factor composition of a spatial visualization test. *Educational and Psychological Measurement*, *14*, 106–119.

Author Index

Subject Index

MIX
Papier aus verantwortungsvollen Quellen
Paper from responsible sources
FSC® C105338

If you have any concerns about our products,
you can contact us on
ProductSafety@springernature.com

In case Publisher is established outside the EU,
the EU authorized representative is:
Springer Nature Customer Service Center GmbH
Europaplatz 3, 69115 Heidelberg, Germany

Printed by Libri Plureos GmbH
in Hamburg, Germany